ENVIRONMENTAL HEALTH - PHYSICAL, CHEMICAL AND BIOLOGICAL FACTORS

COAL MINE DUST

HEALTH REVIEWS AND KEY STUDIES (WITH ACCOMPANYING CD)

ENVIRONMENTAL HEALTH - PHYSICAL, CHEMICAL AND BIOLOGICAL FACTORS

Additional books in this series can be found on Nova's website under the Series tab.

Additional E-books in this series can be found on Nova's website under the E-book tab.

PUBLIC HEALTH IN THE 21ST CENTURY

Additional books in this series can be found on Nova's website under the Series tab.

Additional E-books in this series can be found on Nova's website under the E-book tab.

ENVIRONMENTAL HEALTH - PHYSICAL,
CHEMICAL AND BIOLOGICAL FACTORS

COAL MINE DUST

HEALTH REVIEWS AND KEY STUDIES (WITH ACCOMPANYING CD)

PATRICK FITZHUGH
AND
EILEEN MARGACH
EDITORS

New York

For permission to use material from this book please contact us:
Telephone 631-231-7269; Fax 631-231-8175
Web Site: http://www.novapublishers.com

LIBRARY OF CONGRESS CATALOGING-IN-PUBLICATION DATA

ISBN: 978-1-62417-097-3

Published by Nova Science Publishers, Inc. † New York

CONTENTS

PREFACE

Coal mine dust is one of the most serious occupational hazards in the coal mining industry, and overexposure can cause coal workers' pneumoconiosis (CWP) and a number of other lung diseases, collectively referred to a black lung disease. CWP has been the underlying or contributing cause of death for more than 75,000 coal miners since 1968. This book provides an overview of key studies, background history and health reviews on occupational exposure to respirable coal mine dust.

Chapter 1 - Coal mine dust is one of the most serious occupational hazards in the coal mining industry, and overexposure can cause coal workers' pneumoconiosis (CWP) and a number of other lung diseases, collectively referred to as black lung disease.[1] CWP has been the underlying or contributing cause of death for more than 75,000 coal miners since 1968, according to the Department of Health and Human Services' (HHS) National Institute for Occupational Safety and Health (NIOSH), the federal agency responsible for conducting research on work-related diseases and injuries and recommending occupational safety and health standards. Since 1970, the Department of Labor (Labor) has paid over $44 billion in benefits to miners totally disabled by respiratory diseases (or their survivors), including CWP, through the Black Lung Benefits Program.

Chapter 2 – Since its inception in 1970 the National Institute for Occupational Safety and Health (NIOSH) has extensively investigated and assessed coal miner morbidity and mortality. This history of research encompasses epidemiology; medical surveillance; laboratory-based toxicology, biochemistry, physiology, and pathology; exposure assessment; disease prevention approaches; and methods development. The experience gained in those activities, together with knowledge from external publications and reports, was brought together in 1995 in a major NIOSH review and report of recommendations, entitled *Criteria for a Recommended Standard—Occupational Exposure to Respirable Coal Mine Dust*. This document had the following major recommendations:

1. Exposures to respirable coal mine dust should be limited to 1 mg/m^3 as a time-weighted average concentration for up to a 10 hour day during a 40 hour work week;
2. Exposures to respirable crystalline silica should be limited to 0.05 mg/m^3 as a time-weighted average concentration for up to a 10 hour day during a 40 hour work week;
3. The periodic medical examination for coal miners should include spirometry;
4. Periodic medical examinations should include a standardized respiratory symptom questionnaire;
5. Surface coal miners should be added to and included in the periodic medical monitoring.

This Current Intelligence Bulletin (CIB) updates the information on coal mine dust exposures and associated health effects from 1995 to the present. A principal intent is to determine whether the 1995 recommendations remain valid in the light of the new findings, and whether they need to be updated or supplemented. The report does not deal with issues of sampling and analytical feasibility nor technical feasibility in achieving compliance.

In: Coal Mine Dust
Editors: P. Fitzhugh and E. Margach
ISBN: 978-1-62417-097-3

Chapter 1

MINE SAFETY: REPORTS AND KEY STUDIES SUPPORT THE SCIENTIFIC CONCLUSIONS UNDERLYING THE PROPOSED EXPOSURE LIMIT FOR RESPIRABLE COAL MINE DUST*

United States Government Accountability Office

Coal mine dust is one of the most serious occupational hazards in the coal mining industry, and overexposure can cause coal workers' pneumoconiosis (CWP) and a number of other lung diseases, collectively referred to as black lung disease.[1] CWP has been the underlying or contributing cause of death for more than 75,000 coal miners since 1968, according to the Department of Health and Human Services' (HHS) National Institute for Occupational Safety and Health (NIOSH), the federal agency responsible for conducting research on work-related diseases and injuries and recommending occupational safety and health standards. Since 1970, the Department of Labor (Labor) has paid over $44 billion in benefits to miners totally disabled by respiratory diseases (or their survivors), including CWP, through the Black Lung Benefits Program.

In October 2010, Labor's Mine Safety and Health Administration (MSHA)—the federal agency responsible for setting and enforcing mine safety and health standards—proposed revising the existing standard for coal

* This is an edited, reformatted and augmented version of the Congressional Committees, DC 20548, dated August 17, 2012.

mine dust to lower the permissible exposure limit (PEL)[2] from 2.0 milligrams of dust per cubic meter of air (mg/m^3) to 1.0 mg/m^3.[3] Several coal mining companies and others have questioned the evidence and analytical methods used to support the proposed PEL. In the Consolidated Appropriations Act, 2012, Congress required that GAO review and report on the data collection, sampling methods, and analyses MSHA used to support its proposal.[4] Although MSHA's proposed rule includes other provisions, this review focuses on MSHA's proposal to lower the PEL for coal mine dust from 2.0 mg/m^3 to 1.0 mg/m^3. To respond to this requirement, we addressed the following question: What are the strengths and limitations of the data and analytical methods MSHA used to support its proposal to lower the PEL for coal mine dust?

To conduct our work, we reviewed relevant federal laws and regulations, MSHA's proposed standard to lower the PEL for coal mine dust, the reports and key scientific studies cited in them that MSHA used to support its proposed standard, and the comments MSHA received on its proposal after it was published in the Federal Register. MSHA primarily relied on two reports and the studies cited in them to develop its proposed standard: NIOSH's 1995 *Criteria for a Recommended Standard – Occupational Exposure to Respirable Coal Mine Dust* (Criteria Document) and MSHA's 2010 *Quantitative Risk Assessment in Support of the Proposed Respirable Coal Mine Dust Rule* (Quantitative Risk Assessment). MSHA also relied on the 1996 *Report of the Secretary of Labor's Advisory Committee* (consisting of labor, industry and government representatives) *on the Elimination of Pneumoconiosis Among Coal Mine Workers.*[5] Enclosure I lists and provides a detailed discussion of the reports and key scientific studies we reviewed. The focus of our work was primarily limited to determining whether the scientific studies MSHA used generally support its conclusion that lowering the exposure to coal mine dust would lower miners' risk of disease over their working lives.[6] We assessed the adequacy of the data and measures employed, the reasonableness and rigor of the statistical techniques used to analyze them, and the validity of the conclusions drawn from the analyses. Our work was not designed to determine the optimal PEL for coal mine dust; analyze the costs and benefits of the proposed standard; or determine whether the proposed standard would meet MSHA's legal requirements under the Federal Mine Safety and Health Act of 1977 (Mine Act) or other federal laws that govern the rulemaking process, such as the Administrative Procedure Act.[7]

In addition to reviewing the reports and key scientific studies cited in them that MSHA used to support its proposed standard, we conducted a literature search to identify other studies that examined the relationship between exposure to coal mine dust and its associated health effects. These included an April 2011 NIOSH report on a review of information since 1995 on coal mine dust exposures and associated health outcomes.

Finally, we interviewed MSHA and NIOSH officials and representatives from the mining industry and mine workers. We also reviewed related reports by GAO, Labor's Office of Inspector General, and others on the health effects of exposure to coal mine dust. We visited MSHA and NIOSH offices in Pittsburgh, Pennsylvania and Morgantown, West Virginia, which informed our understanding of the data and the analytical methods used to support MSHA's proposal for lowering the PEL for coal mine dust. A further discussion of our scope and methodology is provided in Enclosure I.

We conducted this performance audit from February 2012 through August 2012 in accordance with generally accepted government auditing standards. These standards require that we plan and perform the audit to obtain sufficient, appropriate evidence to provide a reasonable basis for our findings and conclusions based on our audit objectives. We believe that the evidence obtained provides a reasonable basis for our findings and conclusions based on our audit objectives.

RESULTS IN BRIEF

Our evaluation of the reports MSHA used to support its proposal and the key scientific studies on which the reports were based shows that they support the conclusion that lowering the PEL from 2.0 mg/m^3 to 1.0 mg/m^3 would reduce miners' risk of disease. The reports and key studies concluded that miners' cumulative exposure to coal mine dust at the current PEL over their working lives places them at an increased risk of developing CWP, progressive massive fibrosis, and decreased lung function, among other adverse health outcomes. To mitigate the limitations and biases in the data, the researchers took reasonable steps, such as using multiple x-ray specialists to reduce the risk of misclassifying disease and making adjustments to coal mine dust samples where bias was suspected. In addition to addressing the limitations and biases in the data, researchers used appropriate analytical methods to conclude that lowering the existing PEL would decrease miners' risk of developing black lung disease. For example, in addition to taking steps to precisely estimate a miner's cumulative exposure, the researchers accounted for several factors in their analyses—such as the age of the miners, the carbon content of the coal (coal rank), and other factors known to be associated with the disease—to better estimate the effect of cumulative exposure to coal mine dust.[8] Further, the other studies we identified generally supported the conclusion that reducing the PEL would reduce miners' risk of disease.

BACKGROUND

CWP in the United States

The passage of the Federal Coal Mine Health and Safety Act of 1969 (Coal Act) [9] established the first comprehensive respirable dust standard for coal mines, setting the exposure limit at 2.0 mg/m^3.[10] Following the passage of the Coal Act, the prevalence of CWP among underground coal miners examined in NIOSH's Coal Workers' X-ray Surveillance Program generally decreased about 80 percent from 1970 to 2009.[11] But, according to NIOSH, despite this overall decrease, the observed prevalence of CWP has risen in recent years.[12] NIOSH based this finding on data collected from its Coal Workers' X-ray Surveillance Program and, as a result, it may not be representative of the total population of coal miners since participation in the program is primarily voluntary.

In 2012, coal mine companies employed a total of 86,195 miners in 26 states.[13] While miners across the country are at risk of developing CWP, CWP-related deaths are clustered in the Appalachian region, and, according to NIOSH, clusters of rapidly progressing CWP have been recently observed in Kentucky, Virginia, and West Virginia.[14]

Types of Lung Disease Resulting from Exposure to Coal Mine Dust

Inhaling excessive amounts of coal mine dust can cause CWP and other debilitating lung diseases, including chronic obstructive pulmonary disease, which encompasses chronic bronchitis and emphysema. According to NIOSH, it usually takes about 10 to 15 years of exposure to coal mine dust to develop CWP, although cases involving fewer years of exposure have been observed. Once contracted, CWP cannot be cured, making it critical to prevent the development of this disease by limiting miners' exposure to coal mine dust. Clinical diagnosis of CWP in an individual patient is generally based on the presence of typical chest radiological findings, history of working in coal mines, and exclusion of alternative diagnoses. Although CWP in its early stages may not be associated with impaired lung function or increased mortality, it can increase a miner's risk of developing the advanced stage of the disease, known as progressive massive fibrosis, which can significantly decrease lung function and result in death. According to NIOSH, several factors are associated with a higher risk of developing CWP, including miners' level of exposure to coal mine dust, coal rank, their length of employment in mining (especially years worked underground), their age, and their occupations in the mines.[15]

Role of MSHA

MSHA is responsible for protecting miners by enforcing the provisions of the Mine Act, as amended by the Mine Improvement and New Emergency Response Act of 2006.[16] Under these laws, MSHA has a number of responsibilities, including setting new safety and health standards and revising existing standards, approving training programs for mine workers, and developing regulations regarding training requirements for rescue teams, among other things. MSHA also conducts periodic mine inspections and, along with coal mine operators, periodically collects samples of coal mine dust to determine compliance with the PEL. According to MSHA, approximately 750,000 coal mine dust samples have been collected by inspectors and about 4.6 million dust samples have been collected by mine operators since 1970.[17]

MSHA inspectors and mine operators measure the concentration of coal mine dust over an entire production shift, or at a maximum, an 8-hour period, to determine compliance with the current dust standard. MSHA generally determines compliance with the PEL based on the average of coal mine dust concentration samples taken by the mine operator during five consecutive normal production shifts or five normal production shifts worked on consecutive days.[18] Determinations of compliance are also based on an average of multiple measurements taken by an MSHA inspector.

When MSHA sets standards for toxic materials such as coal mine dust, the Mine Act requires the agency to set standards "which most adequately assure on the basis of the best available evidence that no miner will suffer material impairment of health or functional capacity even if such miner has regular exposure to the hazards…for the period of his working life."[19] In developing a standard, the Mine Act also requires MSHA to consider, among other factors, the feasibility of the standard, and MSHA conducts analyses to determine whether a proposed standard is both economically and technologically feasible.[20] Specific to coal mine dust, the Mine Act further specifies that one of its purposes is to "provide, to the greatest extent possible, that the working conditions in each underground coal mine are sufficiently free of respirable dust concentrations…to permit each miner the opportunity to work underground during the period of his entire adult working life without incurring any disability from…[an] occupation-related disease during or at the end of such period."[21]

Role of NIOSH

NIOSH shares some responsibility with MSHA for improving mine safety and protecting miners' health. It conducts research on the causes of work-related diseases and injuries; researches, develops, and tests new technologies and equipment designed to improve mine safety; and recommends occupational safety and health standards, such as the PEL for coal mine dust. NIOSH recommends safety and health standards to MSHA and other regulatory agencies through its guidance documents (e.g., criteria documents), which provide the scientific basis for its recommended standards and a critical review of the scientific and technical information available on the prevalence of hazards, among other information.

NIOSH also administers the Coal Workers' X-ray Surveillance Program, a medical monitoring and surveillance program to detect and prevent lung disease.[22] This program requires mine operators to provide up to three initial chest x-rays for coal miners within specified time frames after their employment begins, followed by voluntary periodic chest x-rays approximately every 5 years thereafter. NIOSH uses this program for disease surveillance, which includes tracking trends, setting prevention and intervention priorities, and assessing prevention and intervention efforts. Miners' chest x-rays are read and classified by at least two x-ray specialists who must meet certain qualifications.[23]

To estimate the prevalence of lung disease among underground coal miners and to study the relationship between miners' lung disease and their level of exposure to coal mine dust, NIOSH developed the National Study of Coal Workers' Pneumoconiosis (NSCWP). Through the NSCWP, NIOSH analyzed epidemiological data for a sample of mines and miners across all major coalfields. The data included miners' chest x-ray findings, lung function test results, and occupational and smoking histories, as well as available results of coal mine dust sampling.[24] According to NIOSH, epidemiological studies examining the relationship between coal mine dust and disease must contain a sufficiently large body of data over a time period that is adequate to derive reliable findings. (CWP generally takes at least 10 years from first exposure, and usually longer, to become clinically apparent).

Reports and Key Scientific Studies MSHA Used to Support Its Proposal Showed That Lowering the PEL Would Reduce Disease Risk

The two primary reports MSHA used to support its proposed standard—NIOSH's Criteria Document and MSHA's Quantitative Risk Assessment—and the six key scientific studies on which those reports were based each concluded that lowering the PEL from 2.0 mg/m^3 to 1.0 mg/m^3 would reduce coal miners' risk of developing disease.[25] Furthermore, these studies concluded that cumulative exposure to coal mine dust over a working life at the current PEL is associated with adverse health outcomes. These outcomes include CWP, progressive massive fibrosis, emphysema, decreased lung function, and mortality. A statistical model used in both NIOSH's Criteria Document and MSHA's Quantitative Risk Assessment showed that reducing exposure to coal mine dust from 2.0 mg/m^3 to 1.0 mg/m^3 over a miner's working life would decrease the risk of developing CWP from 34 percent to 12 percent for miners working in mines with high coal rank (high carbon content), and from 17 percent to 6 percent for miners working in mines with medium- and low-rank coal (lower carbon content).[26] Key scientific studies also showed that reducing miners' cumulative exposure to coal mine dust over their working lives would decrease their risk of developing other respiratory diseases. For example, another model NIOSH used in its Criteria Document showed that the same reduction in exposure to coal mine dust would reduce the risk of a miner developing a dust-caused reduction in lung function by roughly half, and this would be the case for miners across different regions of the United States. However, the studies also concluded that, even if coal mine dust concentrations were successfully reduced to the proposed PEL of 1.0 mg/m^3, miners would still be at some risk of developing disease.[27]

Reasonable Steps Were Taken to Mitigate the Limitations and Biases in the Data MSHA Relied on to Estimate the Health Effects of Exposure to Coal Mine Dust

Researchers who prepared the reports and key scientific studies on which they were based took reasonable steps to mitigate the limitations and biases in the data MSHA relied on to support its proposal to lower the PEL. These steps included:

- *Using a sufficiently large number of coal mine dust samples.* An accurate examination of the association between exposure to coal mine dust and disease depends, in part, on the reliability of the coal mine dust samples. Precisely measuring coal mine dust is difficult, partly because of the large degree of variation in the levels of coal mine dust in a particular area. However, the reports and the key scientific studies on which they were based relied on a sufficiently large number of coal mine dust samples, which helped mitigate the effect of random variation between individual samples and ensure more accurate measurements of coal mine dust overall.[28]
- *Adjusting coal mine dust samples where bias was suspected.* To improve the accuracy of coal mine dust samples, researchers developed specific procedures to adjust the samples where bias was suspected. For example, according to MSHA, coal mine dust samples collected by mine operators have historically been systematically lower than coal mine dust samples collected by MSHA inspectors at the same mines. Because of this downward bias, according to NIOSH, samples taken by operators likely underestimated miners' exposure to coal mine dust. To address this bias, researchers of some of the key scientific studies systematically adjusted coal mine dust samples upward to complete their analyses.
- *Addressing potential limitations where there were lower participation rates.* Researchers also addressed potential limitations in the data resulting from lower participation rates of miners in medical examinations. Although participation of coal miners in some rounds of the medical examinations was lower than in other rounds, researchers conducted additional analyses to check whether those who participated were systematically different from those who did not participate, which could bias the data. For example, chest x-rays used to identify and classify the presence of CWP were obtained from

medical surveys from NIOSH's NSCWP, which was conducted in multiple rounds over roughly two decades.[29] The participation rates for these four rounds of surveys ranged from as low as 52 percent to as high as 90 percent.[30] Low participation in the medical surveys could have introduced bias into the analysis if the characteristics of the miners who did not participate—such as their age, the number of years they had worked in the mines, or other factors associated with the development of disease—differed substantially from those who did participate in the surveys. To mitigate this possible bias, researchers conducted additional analyses of the NSCWP data and found that those who participated in these surveys were similar to those who did not participate in terms of age, years working in underground mines, and presence of CWP in a prior round of the NSCWP.[31] These analyses lend greater confidence to the conclusion that lower participation rates did not influence the results of the studies.

- *Limiting the risk of misclassifying disease.* To further improve the reliability of the medical data used in the key scientific studies, some researchers took steps to limit the risk of misclassifying disease by using x-ray data that were read and classified by multiple x-ray specialists using a standardized process. MSHA largely relied on two key scientific studies to examine the relationship between exposure to coal mine dust and CWP. In one of these studies, researchers used x-ray data classified by one specialist, and in the other study, researchers used x-ray data independently classified by three specialists. Using x-ray data classified by multiple independent specialists limits the possible error that can be introduced when relying on classifications by one specialist. Although one of the two studies used only one x-ray specialist to classify disease, the findings of that specialist were similar to those in the study that used three x-ray specialists to classify disease.[32] In addition, these three x-ray specialists used the same standardized classification system for determining the presence and severity of CWP, consistent with standard practices for epidemiological research.[33] Furthermore, since classifications of disease have been known to vary among multiple x-ray specialists interpreting the same x-ray, researchers mitigated this concern by using selected specialists who were known, based on historical data, to assign CWP classifications at rates close to the median of multiple readers.

The statistical models in the key scientific studies generally included samples of coal mine dust collected from 1968 to 1987, and the researchers used these data to estimate the relationship between exposure to coal mine dust and risk of disease. We find it reasonable that NIOSH and MSHA used these models to estimate risk of disease due to exposure to coal mine dust in present mining conditions. It is possible, however, that changes in the composition of coal mine dust or characteristics of the coal mine workforce since 1987 have altered the relationship between exposure to coal mine dust and risk of disease. For example, the research we reviewed showed that the quartz (silica) component of coal mine dust, which is known to increase the risk of disease, may have increased in certain states in the years since key scientific studies were conducted.[34] It follows that the health risks of a given level of exposure to coal mine dust may be even greater today than estimated in the models. Therefore, we concluded that an increase in silica levels would not undermine the proposal to lower the exposure limit, and that it was reasonable to rely on the coal mine dust samples used in the key scientific studies to estimate disease risk in support of the proposal.[35]

Appropriate Analytical Methods Were Used in the Key Scientific Studies MSHA Relied on to Develop its Proposed Standard

The researchers for the reports and six key scientific studies that supported MSHA's proposed standard used appropriate analytical methods to conclude that lowering miners' exposure to coal mine dust over their working lives reduces their risk of developing black lung disease. For example, they accounted for known factors associated with black lung disease and precisely estimated miners' cumulative exposure to coal mine dust. Previous research established that the development of black lung disease is affected by certain factors other than cumulative exposure to coal mine dust, such as coal rank and a miner's age. By taking these factors into account, the researchers were able to better isolate the effect of cumulative exposure to coal mine dust and provide greater confidence in their assessment of the relationship between exposure to coal mine dust and the development of disease. For example, one key scientific study that examined the association between lung function and exposure to coal mine dust accounted for each miner's age, height, cigarette smoking, ethnicity, mining status (current or ex-miner), and years worked in non-mining yet dusty occupations. In all of the key scientific studies, coal mine dust was found to be a statistically significant predictor of health risk, even when taking other known factors into account.[36] Furthermore, some key scientific studies tested several different combinations of factors that may contribute to disease to further assess the effect of each factor and better determine the relationship between exposure to coal mine dust and development of disease.[37] While many known factors were taken into account, none of the key scientific studies accounted for exposure to silica dust because, according to NIOSH, reliable information on silica exposure was not available at the time these studies were conducted.

In addition to accounting for known factors associated with disease to better isolate the effect of exposure to coal mine dust on health outcomes, researchers took steps to increase the precision of a miner's estimated cumulative exposure to help ensure greater reliability in their predictions of health outcomes. The key scientific studies estimated each miner's cumulative exposure to coal mine dust from the beginning of his or her employment as a miner to the time of his or her medical examination. Estimating the cumulative exposure to coal mine dust over many years is important because, according to NIOSH, black lung disease generally develops slowly over time. Steps were taken to collect coal mine dust measurements in specific mines and include specific occupations for each year of the study so that precise estimates of a coal miner's exposure could be calculated based on the miner's actual occupation and tenure in that occupation.[38] Some key scientific studies included miners who began mining prior to the establishment of the first PEL in 1970 and MSHA's systematic sampling of coal mine dust concentrations.[39] For these miners, the researchers took reasonable steps to estimate miners' exposure to coal mine dust in years prior to 1970, despite having limited data for those years. Furthermore, to ensure that the conclusions did not depend on the exact method used to estimate coal mine dust exposure prior to 1970, the researchers used various alternatives to estimate a miner's exposure to coal mine dust. Using these alternative methods, the researchers found similar results, lending greater confidence in the accuracy of the estimates of disease risk.

While the analytical methods researchers used in the six key scientific studies supported the conclusion that lowering coal mine dust exposure would reduce the risk of disease for miners over their working lives, the actual estimates of disease risk for miners at the end of their working lives may lack precision. The Mine Act requires MSHA to set standards to protect miners even if they are exposed to hazards for their full working lives. Therefore, predicting the risk of disease at the end of a miner's career (for example, miners age 65 with 45 years of employment) is important when considering a new PEL. However, in most of the key scientific studies, few miners actually had 45 or more years work experience in a mine, and few were 65 or older. Because the key scientific studies predicted disease risk at the end of a miner's career, the estimated levels of risk may not be as precise as if there had been more data on miners at the end of their careers. This limitation, however, does not undermine the studies' overall conclusion that a reduction in exposure to coal mine dust is associated with a reduction in the risk of disease.

In addition to our evaluation of the reports MSHA used to support its proposal and the key scientific studies on which the reports were based, we identified and reviewed several other scientific studies that examined the association between exposure to coal mine dust and respiratory disease. See Enclosure I for additional information on these other studies. These other studies generally supported the conclusion that lowering miners' exposure to coal mine dust would reduce their risk of developing CWP and other dust-induced respiratory diseases. We also found no evidence of other findings or other methodological approaches that would call into question the underlying conclusions in the key scientific studies on which MSHA based its proposal.

AGENCY COMMENTS AND OUR EVALUATION

We provided a draft of this report to Labor and HHS to obtain their comments. In its written comments, Labor agreed with our findings. HHS also agreed with our findings, but did not provide formal written comments. Labor and HHS both provided technical comments, which we incorporated as appropriate.

We are sending copies of this report to the appropriate congressional committees and the Secretaries of Labor and Health and Human Services.

Revae Moran
Director
Education, Workforce, and Income Security Issues

List of Committees

The Honorable Tom Harkin
Chairman
The Honorable Richard C. Shelby

Ranking Member
Subcommittee on Labor, Health and Human Services,
Education, and Related Agencies
Committee on Appropriations
United States Senate

The Honorable Denny Rehberg
Chairman
The Honorable Rosa L. DeLauro
Ranking Member
Subcommittee on Labor, Health and Human Services,
Education, and Related Agencies
Committee on Appropriations
House of Representatives

ENCLOSURE I. SCOPE AND METHODOLOGY

In the Consolidated Appropriations Act, 2012, Congress required that GAO review and report on the data collection, sampling methods, and analyses the Department of Labor's (Labor) Mine Safety and Health Administration (MSHA) used to support its recent proposal to lower the permissible exposure limit (PEL) for respirable coal mine dust[40] from 2.0 milligrams of dust per cubic meter of air (mg/m^3) to 1.0 mg/m^3.[41] Although MSHA's proposal includes other provisions,[42] our review focused on MSHA's proposal to lower the PEL. To respond to this requirement, we answered the following question: What are the strengths and limitations of the data and analytical methods MSHA used to support its proposal to lower the permissible exposure limit for coal mine dust?

To conduct our work, we reviewed relevant federal laws and regulations, MSHA's proposed standard for exposure to coal mine dust, the comments MSHA received in response to the proposed rule, and the reports and the key scientific studies MSHA used to support its proposed standard. We also reviewed additional scientific studies as well as related reports by GAO, Labor's Office of Inspector General, and others on the health effects of exposure to coal mine dust. We interviewed officials with MSHA and the Department of Health and Human Services' National Institute for Occupational Safety and Health (NIOSH), representatives from the mining industry and mine workers. We also conducted site visits to MSHA and

NIOSH offices in Pittsburgh, Pennsylvania and Morgantown, West Virginia, which informed our understanding of the data and analytical methods used to support MSHA's proposal to lower the PEL for coal mine dust.

We conducted this performance audit from February 2012 through August 2012 in accordance with generally accepted government auditing standards. These standards require that we plan and perform the audit to obtain sufficient, appropriate evidence to provide a reasonable basis for our findings and conclusions based on our audit objectives. We believe that the evidence obtained provides a reasonable basis for our findings and conclusions based on our audit objectives.

Analysis of Reports Used by MSHA and the Key Scientific Studies on Which They Were Based

To determine the strengths and limitations of the data and analytical methods MSHA used to support its proposal to lower the PEL for coal mine dust, we identified and reviewed the reports and key scientific studies cited in them that MSHA relied on to support its proposed standard. MSHA primarily relied on two reports and the studies cited in them to develop its proposed standard: NIOSH's 1995 *Criteria for a Recommended Standard – Occupational Exposure to Respirable Coal Mine Dust* (Criteria Document) and MSHA's 2010 *Quantitative Risk Assessment in Support of the Proposed Respirable Coal Mine Dust Rule* (Quantitative Risk Assessment). Table 1 lists the reports and the key scientific studies. We selected these reports because they were used by MSHA to develop or support the proposed rule.[43] In addition, we reviewed MSHA's policies and procedures for its coal mine dust sampling program and NIOSH's processes for administering the Coal Workers' X-ray Surveillance Program.[44] The focus of our work was limited to determining whether the scientific studies MSHA used generally supported its conclusion that lowering the PEL from 2.0 mg/m^3 to 1.0 mg/m^3 would lower the risk of disease over a miner's working life.[45] We assessed the adequacy of the data and measures employed, the reasonableness and rigor of the statistical techniques used to analyze them, and the validity of the conclusions drawn from the analyses. Our work was not designed to determine the optimal limit for coal mine dust exposure; assess the cost-benefit analysis MSHA conducted in support of its proposed standard; or determine whether MSHA's proposed rule complied with any legal requirements imposed by the Federal Mine Safety and Health Act of 1977 (Mine Act), the Administrative Procedure Act, or other applicable federal laws or executive orders that govern the rulemaking process.[46]

Table 1. Reports and Key Scientific Studies MSHA Used to Support Its Proposed Standard

Report	Key Studies Cited in the Report	Description of Study
NIOSH's 1995 Criteria for a Recommended Standard: Occupational Exposure to Respirable Coal Mine Dust	Attfield, M.D. and K. Morring, "An Investigation into the Relationship Between Coal Workers' Pneumoconiosis and Dust Exposure in U.S. Coal Miners" (1992)	Provides an estimate of the relationship between the cumulative coal mine dust miners are exposed to over their working lives and disease prevalence in U.S. coal miners.
	Attfield, M.D. and N.S. Seixas, "Prevalence of Pneumoconiosis and its Relationship to Dust Exposure in a Cohort of U.S. Bituminous Coal Miners and Ex-Miners" (1995)	Provides an estimate of the relationship between the cumulative coal mine dust miners are exposed to over their working lives and disease prevalence among current and former miners.
	Attfield M.D. and T.K. Hodous, "Pulmonary Function of U.S. Coal Miners Related to Dust Exposure Estimates" (1992)	Provides an estimate of the effects of coal mine dust exposure on lung function in U.S. coal miners.
	Seixas N.S., T.G. Robins, M.D. Attfield, L.H. Moulton, "Longitudinal and Cross Sectional Analyses of Exposure to Coal Mine Dust and Pulmonary Function in New Miners" (1993)	Provides an estimate of the effects of coal mine dust exposure on lung function in U.S. coal miners.
MSHA's 2010 Quantitative Risk Assessment	Attfield, M.D. and N.S. Seixas, "Prevalence of Pneumoconiosis and its Relationship to Dust Exposure in a Cohort of U.S. Bituminous Coal Miners and Ex-Miners" (1995)[a]	Provides an estimate of the relationship between the cumulative coal mine dust miners are exposed to over their working lives and disease prevalence among current and former miners.
	Kuempel et al., "Emphysema and Pulmonary Impairment in Coal Miners: Quantitative Relationship with Dust Exposure and Cigarette Smoking" (2009)	Provides predictions of the relationship between the cumulative coal mine dust miners are exposed to and their risk of developing emphysema.
	Attfield, M.D. and E.D. Kuempel, "Mortality Among U.S. Underground Coal Miners: A 23-Year Follow-Up" (2008)	Examines the contributing and underlying causes of death in U.S. coal miners.

Source: GAO review of reports.

[a] Attfield, M.D., and N.S. Seixas (1995) is a key study that was used in both the NIOSH Criteria Document and the MSHA Quantitative Risk Assessment.

Through interviews with NIOSH officials, we identified four key scientific studies in its Criteria Document that served as the primary basis for its recommendation to lower the PEL from 2.0 mg/m^3 to 1.0 mg/m^3. During our review of the Criteria Document, we identified and reviewed an additional five studies that contained important details necessary for evaluating NIOSH's report.[47] In addition, through an interview with the principal author of the Quantitative Risk Assessment and MSHA officials, we identified three key scientific studies used in that report to quantify the relationship between coal mine dust exposure and disease risk. Two GAO specialists with expertise in social science methods, statistics, or epidemiology and public health examined each study to assess the adequacy of the samples and measures employed, the reasonableness and rigor of the statistical techniques used to analyze them, and the validity of the conclusions drawn from the analyses. For selected studies, we contacted the researchers directly as necessary for clarification or additional information.

NIOSH's Criteria Document included a review of all available information through 1995 and estimated the adverse health effects associated with exposure to coal mine dust over a miner's working life to provide the scientific basis for NIOSH's recommendation for a reduced PEL. In this report, NIOSH concluded that epidemiological studies demonstrated that miners have an elevated risk of developing black lung disease, which includes coal workers pneumoconiosis (CWP) and other respiratory or pulmonary impairments, when they are exposed over their working lives to coal mine dust at the current PEL of 2.0 mg/m^3. This conclusion represents a change in the understanding of disease risk since the time the 2.0 mg/m^3 PEL was initially established.[48] Further, based on its evaluation of health effects data, the feasibility of collecting and analyzing coal mine dust samples, and technological feasibility of controlling exposures, NIOSH recommended that the PEL be reduced to 1.0 mg/m^3 as a time-weighted average concentration for up to 10 hours per day during a 40-hour work week. NIOSH concluded, however, that even at an exposure level of 1.0 mg/m^3—the lowest PEL NIOSH considered feasible—miners exposed at this concentration over a working life still have a risk of developing black lung disease.

In response to NIOSH's Criteria Document in support of its recommendation to lower the PEL, MSHA completed a Quantitative Risk Assessment in September 2010 to determine whether current conditions involving exposure to coal mine dust place miners at risk for developing black lung and whether the proposed rule will substantially reduce those risks. The report addressed three questions related to MSHA's proposed rule: whether (1) the potential health effects associated with current exposure levels constitute material impairments to a miner's health or functional capacity; (2) current conditions place miners at a significant risk of incurring any of these material impairments; and (3) the proposed rule will substantially reduce those risks.[49] MSHA's report concluded that the current exposure level of 2.0 mg/m^3 placed miners at a significant risk of incurring each of the material impairments considered and that the proposed rule would substantially reduce the risks of CWP, severe emphysema, and mortality attributable to coal mine dust exposure.[50]

We also reviewed NIOSH's April 2011 report, *Coal Mine Dust Exposures and Associated Health Outcomes: A Review of Information Published Since 1995*.[51] This report provides a summary of information on exposure to coal mine dust and its associated health effects since 1995. NIOSH's intent was to determine whether its recommendations in 1995 to lower the PEL to 1.0 mg/m^3 remained valid in light of new findings. We reviewed the sections of the report most relevant to respiratory disease outcomes. We also reviewed two studies that were referred to in this report that were especially relevant to our research objective, as they contained additional information about the relationship between exposure to coal mine dust and risk of disease in British coal miners.[52]

Additional Scientific Studies Reviewed

We also conducted a search for studies that explored the relationship between exposure to coal mine dust and respiratory disease. This search drew on several sources:

- an extensive literature search for such studies;
- a review of materials submitted to MSHA in response to its proposed rule during the public comment period; and,
- interviews with MSHA and NIOSH officials, mining company representatives, and representatives of mine workers.

The initial search netted hundreds of research studies. We applied the following criteria in order to limit our review to include only studies with the potential to shed new light on the scientific support for the proposal to lower the PEL:

(1) studies not cited in the two reports MSHA used to develop its proposed rule, or any of the key scientific studies on which they were based;[53]
(2) studies published in the last 50 years and before October 19, 2010, the publication date of MSHA's proposed rule; and,
(3) studies containing original analysis (as opposed to those that used or relied on research already conducted).

This effort yielded 10 research studies that met our criteria (see table 2).

Table 2. Additional Studies That Met Our Criteria

(1) Leigh, J., Todorovic, M., & Driscoll, T., "Histological Changes in Hilar Lymph Glands in Relation to Coal Workers' Progressive Massive Fibrosis and Lung Coal and Quartz Contents," *The Annals of Occupational Hygiene*, Vol.41, (1997)
(2) Wang, X.R. and D. Christiani, "Respiratory Symptoms and Functional Status in Workers Exposed to Silica, Asbestos, and Coal Mine Dusts," *Journal of Occupational and Environmental Medicine*, Vol.42, No.11, (2000)
(3) Noble, R.B., Bailer, A.J., & Park, R., "Model-Averaged Benchmark Concentration Estimates for Continuous Response Data Arising from Epidemiological Studies," *Risk Analysis*, Vol.29, No.4, (2009)
(4) Liu, H., Tang, Z., Yang, Y., Weng, D., Sun, G., Duan, Z., & Chen, J., "Identification and Classification of High Risk Groups for Coal Workers' Pneumoconiosis Using an Artificial Neural Network Based on Occupational Histories: A retrospective Cohort Study," *BMC Public Health*, Vol.9, No.366, (2009)
(5) Liu, H., Tang, Z., Weng, D., Yang, Y., Tian, L., Duan, Z., & Chen, J., "Prevalence Characteristics and Prediction of Coal Workers' Pneumoconiosis in the Tiefa Colliery in China," *Industrial Health*, Vol.47, (2009)
(6) Ashford, J.R., Fay, J.W.J., & Smith, C.S., "The Correlation of Dust Exposure with Progression of Radiological Pneumoconiosis in British Coal Miners," *American Industrial Hygiene Association Journal*, Vol.26, No.4, (1965)
(7) Mamuya, S. HD., Bratveit, M., Mashalla, Y., & Moen, B., "High Prevalence of respiratory Symptoms Among Workers in the Development Section of a Manually Operated Coal Mine in a Developing Country: A Cross Sectional Study," *BMC Public Health*, Vol.7, No.17, (2007)
(8) Atlin, R., Savranlar, A., Kart, L., Mahmutyazicioglu, K., Ozdemir, H., Akdag, B., & Gundogdu, S., "Presence and HRCT Quantification of Bronchiectasis in Coal Workers," *European Journal of Radiology*, Vol,52, (2004)
(9) Morefeld, P., Ambrosy, J., Bengtsson, U., Bicker, H., Kalkowsky, B., Kosters, A., Lenaerts, H., Ruther, M., Vautrin, H.J., & Piekarski, C., "The Risk of Developing Coal Workers' Pneumoconiosis in German Coal Mining Under Modern Mining Conditions," *The Annals of Occupational Hygiene*, Vol.46, (2002)
(10) Attfield, M.D., Kuempel, E., & Wagner, G., "Exposure Response for Coal Workers' Pneumoconiosis in Underground Coal Miners: A Discussion of Issues and Findings," *The Annals of Occupational Hygiene*, Vol.41, (1997)

Source: GAO review of additional studies.

Each of these studies was then reviewed by a GAO specialist with expertise in social science methods, statistics, or epidemiology and public health. Each specialist made an initial determination of whether the study's results were inconsistent with, or revealed limitations in the work used to support the proposal to lower the PEL. If a study met this criterion, then an additional GAO specialist conducted a further review to assess the adequacy of the samples and measures employed, the reasonableness and rigor of the statistical techniques used to analyze them, and the validity of the conclusions drawn from the analyses.

We based our findings primarily on the results of the models used in the key scientific studies, which are based on data for U.S. miners only. While our methodology included a review of other studies, we found only one different type of model employed by others that attempted to estimate disease risk based on dust exposure.[54] This model used data on British coal miners only. We did not evaluate whether the predicted estimates of disease risk in the key scientific studies of U.S. coal miners would be similar if they were generated from the models presented in other studies. The key scientific studies were used to determine the predicted risk of disease due to exposure to coal mine dust, according to NIOSH, because they are more relevant to mining conditions in the United States. The results from other studies generally showed lower risk of CWP and progressive massive fibrosis compared with the predicted risk of disease in the key scientific studies. These differences in predicted risk, however, may be due to a number of factors, including differences in the models and the underlying assumptions used. While these differences may exist, they do not indicate that the relationship between coal mine dust exposure and disease risk is different in different countries. For example, all of the studies we reviewed indicate that a reduction in exposure to coal dust is associated with a reduction in risk of developing disease.

Interviews with MSHA, NIOSH, and Other Groups

To further inform our review of the evidence used to develop the proposed rule, we interviewed MSHA and NIOSH officials as well as representatives from the mining industry and mine workers. At MSHA, we spoke with officials in the Office of Standards, Regulations, and Variances, the Safety and Health Technology Center, and its District 2 field office in Prosperity, Pennsylvania. At NIOSH, we spoke with officials in its Office of Mine Safety and Health Research in Pittsburgh, Pennsylvania; its Division of Respiratory

Disease Studies in Morgantown, West Virginia; its Education and Information Division in Cincinnati, Ohio; and its Office of the Director in Washington, D.C. We also spoke with several representatives from the mining industry and mine workers, including officials from the National Mining Association, the Bituminous Coal Operators' Association, the United Mine Workers of America, Murray Energy, Alliance Coal Company, and Consol Energy.

Site Visits

To obtain information on the research on improved dust control technologies and sampling methods used to measure the level of coal mine dust to which miners are exposed, we visited NIOSH's Office of Mine Safety and Health Research in Pittsburgh, Pennsylvania. To obtain information about epidemiological studies used to support NIOSH's recommendation to lower the existing PEL for coal mine dust, we also visited NIOSH's Division of Respiratory Disease Studies, in Morgantown, West Virginia. Officials from NIOSH's Education and Information Division, in Cincinnati, Ohio, and NIOSH's Office of the Director, in Washington, D.C. also participated in these discussions. In addition, we visited MSHA's Safety and Health Technology Center at the Bruceton Laboratory in Pittsburgh, Pennsylvania to obtain information on how samples containing coal mine dust, including silica, are processed and analyzed. To observe the conditions under which underground coal miners work, we also visited the Bailey Coal Mine in Wind Ridge, Pennsylvania.

End Notes

[1] In this report, we use the term coal mine dust to refer to respirable coal mine dust. Black lung is a term that includes CWP and other chronic respiratory or pulmonary impairments resulting from coal mine employment.

[2] In this report, the term PEL refers to the respirable coal mine dust standard.

[3] Lowering Miners' Exposure to Respirable Coal Mine Dust, Including Continuous Personal Dust Monitors, 75 Fed. Reg. 64,412 (Oct. 19, 2010) (to be codified at 30 C.F.R. pts. 70, 71, 72, 75, and 90). As of August 13, 2012, MSHA had not finalized this regulation.

[4] The Act provided that "[n]one of the funds made available by this Act may be used to implement or enforce [MSHA's] proposed rule" until GAO "issues, at a minimum, an interim report…and…not later than 240 days after enactment of this Act, submits the report …to the Committees on Appropriations of the House of Representatives and the Senate," or until that deadline has passed. Pub. L. No. 112-74, div. F, tit. I, § 112, 125 Stat. 786, 1064 (2011).

[5] This report was included in our overall review of relevant materials, but we did not include it in our final analysis of the two reports and the key studies cited in them on which MSHA relied to develop its proposed standard because it did not contain new information or analyses of the relationship between coal mine dust exposure and its associated health effects.

[6] For purposes of this proposed rule, MSHA defines a miner's working life to be 45 years.

[7] Various federal statutes and executive orders require federal agencies to follow a number of procedural and analytic requirements when developing and issuing rules. For example, the Administrative Procedure Act generally requires agencies to publish proposed rules for public comment prior to issuing a final rule. 5 U.S.C. § 553. In addition, section 101(a) of the Mine Act establishes certain procedural and analytic rulemaking requirements for MSHA standards. For example, before issuing a standard on toxic materials, MSHA must determine, based on research and other considerations, that the toxins pose a material impairment to miners' health or functional capacity. 30 U.S.C. § 811(a)(6)(A).

[8] Coal rank is a classification of the amount of carbon in the coal. Coal is typically ranked as "high," "medium" or "low," where coal with a high rank has high carbon content. According to NIOSH, several factors are associated with a higher risk of developing CWP, including miners' exposure to coal mine dust with high carbon content.

[9] Pub. L. No. 91-173, 83 Stat. 742.

[10] The Coal Act set an interim standard of 3.0 mg/m^3 that went into effect in 1970, 6 months after the date of enactment and continued until 1972, when the standard of 2.0 mg/m^3 then took effect. In 1977, the Mine Act was passed, which amended and renamed the Coal Act and established MSHA. Pub. L. No. 95-164, 91 Stat. 1290 (1977). The Mine Act maintained the standard at 2.0 mg/m^3, to be in effect until superseded by improved standards issued by MSHA. The Mine Act also authorized MSHA to issue improved standards, and prohibited it from issuing standards that would reduce the protection provided by existing standards. In 1980, via the rulemaking process, MSHA issued the current standard, keeping the PEL at 2.0 mg/m^3. 45 Fed. Reg. 23,990 (Apr. 8, 1980), codified at 30 C.F.R. §§ 70.100, 71.100.

[11] NIOSH's Coal Workers' X-ray Surveillance Program is a primarily voluntary medical monitoring and surveillance program. NIOSH reports prevalence data as averages of 5-year intervals. The overall decrease since 1970 reported here is based on the oldest and most recent intervals: 1970 to1974 and 2005 to 2009.

[12] According to NIOSH, the prevalence of CWP appears to have stopped declining around the 1995 to 1999 interval and has been on the rise since that time.

[13] According to the Department of Energy's Office of Oil, Gas, and Coal Supply Statistics, Wyoming mines the most coal, followed by West Virginia, Kentucky, and Pennsylvania.

[14] Appalachia includes Alabama, Ohio, Kentucky, Pennsylvania, Tennessee, Virginia, and West Virginia.

[15] Miners work in different occupations and locations within the mine and, as a result, are exposed to different levels of coal mine dust.

[16] Pub. L. No. 109-236, 120 Stat. 493, codified as amended at 30 U.S.C. §§ 801-965.

[17] These figures include only samples for underground mines.

[18] See 30 C.F.R. §§ 70.201(b), 70.207, 70.208 (for underground coal mines). Similar sampling procedures apply to surface mines. 30 C.F.R. §§ 71.201(b), 71.208.

[19] 30 U.S.C. § 811(a)(6)(A).

[20] See *Nat'l Mining Ass'n v. Sec'y of Labor*, 153 F.3d 1264, 1269 (11th Cir. 1998).

[21] 30 U.S.C. § 841(b).

[22] This program is part of NIOSH's Coal Workers' Health Surveillance Program, carried out pursuant to requirements in the Mine Act and NIOSH regulations. 30 U.S.C. § 843; 42 C.F.R. §§ 37.1 - 37.80.

[23] Through NIOSH's B Reader program—a training and testing program that began in 1974—a pool of qualified readers is established, using initial and periodic examinations to verify the competence of physician-readers in assessing and classifying pneumoconiosis.

[24] NIOSH's NSCWP was conducted in four rounds of medical surveys. This NSCWP differs from NIOSH's Coal Workers' X-ray Surveillance Program in that the NSCWP achieved higher participation rates and collected more detailed medical information for each miner than the Coal Workers X-ray Surveillance Program.

[25] See Enclosure I for additional information on these reports and studies.

[26] These estimates are for a hypothetical population of 65-year-old miners who have been exposed to coal mine dust over a 45-year working life.

[27] According to NIOSH, the goal of its recommendation to lower the PEL is to minimize, to the greatest extent possible, the health risks associated with a miner's exposure to coal mine dust. In making this recommendation, NIOSH took into account an evaluation of health effects data as well as the feasibility of collecting and analyzing dust samples. NIOSH also considered the technological feasibility of controlling exposures. Specifically, NIOSH determined that the recommended PEL of 1.0 mg/m^3 would have clear beneficial health effects for all miners. At the time the Criteria Document was published, however, a PEL lower than 1.0 mg/m^3 was not considered to be technologically feasible. According to NIOSH, if improvements in the technological feasibility of dust controls become available in the future, the agency would consider recommending a lower PEL.

[28] To develop the statistical model of CWP used by NIOSH in its Criteria Document and by MSHA in its Quantitative Risk Assessment, researchers drew on 293,292 coal mine dust samples in order to estimate a personal cumulative coal mine dust exposure for each of 3,194 miners.

[29] Medical surveys from the NSCWP included a chest x-ray, test of lung function, and a questionnaire on current symptoms, demographics, smoking and work history. Examinations in Round 1 were conducted from 1969 to 1971, Round 2 from 1972 to 1975, Round 3 from 1977 to 1981, and Round 4 from 1985 to 1988.

[30] Participation rates of eligible participants were 90, 75, 52, and 70 percent for Rounds 1 though 4, respectively.

[31] In Round 4 of the NSCWP—which was the basis for one of the key exposure-response models relating dust exposure and risk of CWP—miners who participated in the study and those who did not were very similar with regard to age. Non-participants, however, had worked underground for a slightly longer period of time. Despite this, the non-participants had a slightly lower prevalence of CWP in an earlier round of the study. None of the key scientific studies we reviewed included data collected from Round 3—the Round with the lowest participation rate (52 percent).

[32] The study that used only one x-ray specialist originally included classifications from three specialists who classified disease, but the classifications of two of them were excluded because one specialist did not apply the correct system of classification and the other specialist's readings were found to show unusually high levels of disease.

[33] These x-ray specialists used the International Labour Organization classification system, an internationally-accepted means for assessing the severity and types of abnormality arising from inhalation of mineral dusts. A summary classification of CWP was used for each participant by taking the median category of disease when three x-ray readings were available. The three specialists classified similar overall prevalences of CWP (7 percent, 7 percent, and 9 percent, respectively).

[34] Some mining involves cutting and extracting rock containing silica overlying or underlying the coal seam. Inhaling silica can cause silicosis, a form of pneumoconiosis. Currently, when the respirable coal mine dust contains more than 5 percent quartz (silica), the PEL is determined based on a formula that reduces the PEL commensurate with the proportion of the dust that is silica. 30 C.F.R. §§ 70.101, 71.101. NIOSH has recommended a separate PEL for silica of 0.05 mg/m^3 and MSHA is planning to establish a separate standard for silica.

[35] Other factors that may have changed since 1987 and could affect the relationship between exposure to coal mine dust and disease risk include the prevalence of smoking among coal

miners and the age composition of the mining workforce. Although smoking has its own health effects, it does not appear to alter the risk of CWP due to exposure to coal mine dust, according to researchers. We did not identify any other research on how age affects the relationship between exposure to coal mine dust and disease.

[36] Specifically, coefficients for coal mine dust exposure were statistically significant in all of the key models used by MSHA and NIOSH. A coefficient is considered statistically significant if the probability of observing a value as large as it is, due to chance alone, is less than a specified probability—often 5 percent. In this case, statistical significance means that the observed association between coal mine dust exposure and disease risk, after accounting for other factors such as age and coal rank, is unlikely to be due to chance alone.

[37] The model fit, or how well the model's predictions of disease prevalence matched observed prevalence, was generally good for the key scientific studies. For example, NIOSH evaluated how well model predictions matched actual observations and found that the model-predicted prevalence of early stages of CWP fit the observed prevalence closely within the range of cumulative dust exposure corresponding to the current and proposed PEL.

[38] Information on a miners' work history, including the dates of starting and stopping work in each occupation in each mine, was obtained from interviews of each participant of the NSCWP.

[39] Prior to 1970, coal mine dust concentrations often exceeded the current PEL of 2.0 mg/m^3, according to NIOSH. There is evidence indicating that dust concentrations were as high as 6.0 mg/m^3 for certain occupations within mines prior to 1970.

[40] In this report, we use the term coal mine dust to refer to respirable coal mine dust.

[41] Pub. L. No. 112-74, div. F, tit. I, § 112, 125 Stat. 786, 1064 (2011). MSHA's proposed rule was published in October 2010. Lowering Miners' Exposure to Respirable Coal Mine Dust, Including Continuous Personal Dust Monitors, 75 Fed. Reg. 64,412 (Oct. 19, 2010) (to be codified at 30 C.F.R. pts. 70, 71, 72, 75, and 90). As of August 13, 2012, MSHA had not finalized this regulation.

[42] For example, the proposed rule also contained provisions that would change dust sampling procedures— including requirements for the use of continuous personal dust monitors—and expand medical surveillance.

[43] These reports are referenced in the proposed rule, 75 Fed. Reg. 64,412, 64,414, 64,469 (Oct. 19, 2010). We also confirmed with MSHA officials that the agency used these reports to develop and support its proposal to lower the PEL.

[44] The Coal Workers' X-ray Surveillance Program is a component of NIOSH's Coal Workers' Health Surveillance Program, a medical monitoring and surveillance program carried out pursuant to requirements in the Mine Act and NIOSH regulations. 30 U.S.C. § 843, 42 C.F.R. §§ 37.1-37.80.

[45] For purposes of this proposed rule, MSHA defines a miner's working life to be 45 years.

[46] Various federal statutes and executive orders require federal agencies to follow a number of procedural and analytic requirements when developing and issuing rules. For example, the Administrative Procedure Act generally requires agencies to publish proposed rules for public comment prior to issuing a final rule. 5 U.S.C. § 553. In addition, section 101(a) of the Mine Act establishes certain procedural and analytic rulemaking requirements for MSHA standards. For example, before issuing a standard on toxic materials, MSHA must determine, based on research and other considerations, that the toxins pose a material impairment to miners' health or functional capacity. 30 U.S.C. § 811(a)(6)(A).

[47] The five additional studies reviewed contained more detailed information on the derivation of dust exposure estimates, the cohort of miners examined, the sampling design, and the overall analysis methods. The five studies include (1) Attfield, M. D., & Morring, K. "The derivation of estimated dust exposures for U.S. coal miners working before 1970," *American Industrial Hygiene Association*, vol. 53, no. 4 (1992); (2) Seixas, N. S., Moulton, L. H., Robins, T. G., Rice, C. H., Attfield, M. D., & Zellers, E. T. "Estimation of

cumulative exposures for the National Study of Coal Workers' Pneumoconiosis," Applied Occupational and Environmental Hygiene, vol.6, no.12 (1991); (3) Seixas, N. S., Robins, T.G., Attfield, M. D., & Moulton, L. H. "Exposure-response relationships for coal mine dust and obstructive lung disease following enactment of the Federal Coal Mine Health and Safety Act of 1969," *American Journal of Industrial Medicine*, vol. 21, no. 5 (1992); (4) Morgan, W. K. C., Burgess, D. B., Jacobson, G., O'Brien, R. J., & Pendergrass, E. P., Reger, R. B., & Shoub, E. P. "The prevalence of coal workers' pneumoconiosis in US coal miners," *Archives of Environmental Health*, vol. 27, no.4 (1973); and (5) Kuempel, E. D., Smith, R. J., Attfield, M. D., & Stayner, L. T. "Risks of occupational respiratory diseases among U.S. coal miners," *Applied Occupational and Environmental Hygiene*, vol. 12, no.12 (1997).

[48] According to NIOSH, the current PEL of 2.0 mg/m^3 for coal mine dust exposure was based primarily on estimates of earlier studies of coal miners in the United Kingdom, where the probability of disease progression was thought to be zero for miners' exposure to coal mine dust at an average concentration of 2.0 mg/m^3 over a 35-year working life.

[49] The Mine Act establishes the criteria for MSHA's Quantitative Risk Assessment, 30 U.S.C. § 811(a)(6)(A). According to MSHA, these three questions are based on court interpretations of similar statutory language in the Occupational Safety and Health Act of 1970.

[50] We did not assess whether MSHA met applicable statutory requirements in issuing this proposed rule. Specifically, we did not evaluate whether the risk posed by coal mine dust at current conditions constitutes a "material impairment" to a miner's health, whether the current conditions create a "significant risk" of impairment, or whether the proposed rule will "substantially reduce" those risks, as those terms are used in the statute and related court decisions. 30 U.S.C. § 811(a)(6)(A); see, for example, *Indus. Union Dep't, AFL-CIO v. Am. Petroleum Inst.*, 448 U.S. 607, 639 (1980).

[51] This report had not been issued at the time MSHA published its proposed rule in October 2010. Our review of studies in this report was limited to those published before October 2010.

[52] These studies included Hurley J.F., and W.M. Maclaren, "Dust-related risks of radiological changes in coalminers over a 40-year working life: Report on work commissioned by NIOSH," *Institute of Occupational Medicine*. (1987) and Soutar C.A., J.F. Hurley, B.G. Miller, H.A. Cowie, D. Buchanan, "Dust concentrations and respiratory risks in coalminers: key risk estimates from the British pneumoconiosis field research." *Occupational and Environmental Medicine*. 61 (2004).

[53] In addition to MSHA's proposed rule published in 2010, we reviewed two prior proposed rules related to coal mine dust published on July 7, 2000, and March 6, 2003, to further identify any additional epidemiological studies examining the relationship between exposure to coal mine dust and respiratory disease. 65 Fed. Reg. 42,122 (July 7, 2000), 68 Fed. Reg. 10,784 (March 6, 2003).

[54] This model was included in Hurley J.F., and W.M. Maclaren, "Dust-related risks of radiological changes in coalminers over a 40-year working life: Report on work commissioned by NIOSH," *Institute of Occupational Medicine*. (1987).

In: Coal Mine Dust
Editors: P. Fitzhugh and E. Margach
ISBN: 978-1-62417-097-3

Chapter 2

COAL MINE DUST EXPOSURES AND ASSOCIATED HEALTH OUTCOMES—A REVIEW*

Department of Health and Human Services

FOREWORD

Since its inception in 1970 the National Institute for Occupational Safety and Health (NIOSH) has extensively investigated and assessed coal miner morbidity and mortality. This history of research encompasses epidemiology; medical surveillance; laboratory-based toxicology, biochemistry, physiology, and pathology; exposure assessment; disease prevention approaches; and methods development. The experience gained in those activities, together with knowledge from external publications and reports, was brought together in 1995 in a major NIOSH review and report of recommendations, entitled *Criteria for a Recommended Standard—Occupational Exposure to Respirable Coal Mine Dust*. This document had the following major recommendations:

1. Exposures to respirable coal mine dust should be limited to 1 mg/m^3 as a time-weighted average concentration for up to a 10 hour day during a 40 hour work week;

* This is an edited, reformatted and augmented version of the Current Intelligence Bulletin 64, DHHS (NIOSH) Publication No. 2011–172, dated April 2011.

2. Exposures to respirable crystalline silica should be limited to 0.05 mg/m^3 as a time-weighted average concentration for up to a 10 hour day during a 40 hour work week;
3. The periodic medical examination for coal miners should include spirometry;
4. Periodic medical examinations should include a standardized respiratory symptom questionnaire;
5. Surface coal miners should be added to and included in the periodic medical monitoring.

This Current Intelligence Bulletin (CIB) updates the information on coal mine dust exposures and associated health effects from 1995 to the present. A principal intent is to determine whether the 1995 recommendations remain valid in the light of the new findings, and whether they need to be updated or supplemented. The report does not deal with issues of sampling and analytical feasibility nor technical feasibility in achieving compliance.

John Howard, MD
Director, National Institute for Occupational Safety and Health
Centers for Disease Control and Prevention

EXECUTIVE SUMMARY

Information relating to occupational pulmonary disease morbidity and mortality of coal miners available up to 1995 was reviewed in the NIOSH publication: *Criteria for a Recommended Standard—Occupational Exposure to Respirable Coal Mine Dust*, or Coal Criteria Document (CCD). This led to the following principal conclusions concerning health effects associated with coal mining:

1. Exposure to coal mine dust causes various pulmonary diseases, including coal workers' pneumoconiosis (CWP) and chronic obstructive pulmonary disease (COPD).
2. Coal miners are also exposed to crystalline silica dust, which causes silicosis, COPD, and other diseases.
3. These lung diseases can bring about impairment, disability and premature death.

This Current Intelligence Bulletin updates the previously published review with respect to findings relevant to the health of U.S. coal miners published since 1995. The main conclusions are:

1. After a long period of declining CWP prevalence, recent surveillance data indicate that the prevalence is rising.
2. Coal miners are developing severe CWP at relatively young ages (<50 years).
3. There is some indication that early development of CWP is being manifested as premature mortality.
4. The above individuals would have been employed all of their working lives in environmental conditions mandated by the 1969 Coal Mine Health and Safety Act.
5. The increase in CWP occurrence appears to be concentrated in hot spots of disease mostly concentrated in the central Appalachian region of southern West Virginia, eastern Kentucky, and western Virginia.
6. The cause of this resurgence in disease is likely multifactorial. Possible explanations include excessive exposure due to increases in coal mine dust levels and duration of exposure (longer working hours), and increases in crystalline silica exposure (see below). As indicated by data on disease prevalence and severity, workers in smaller mines may be at special risk.
7. Given that the more productive seams of coal are being mined out, a transition by the industry to mining thinner coal seams and those with more rock intrusions is taking place and will likely accelerate in the future. Concomitant with this is the likelihood of increased potential for exposure to crystalline silica, and associated increased risk of silicosis, in coal mining.

The main conclusions drawn from review of the new information are:

1. While findings published since 1995 refine or add further to the understanding of the respiratory health effects of coal mine dust described in the NIOSH CCD, they do not contradict or critically modify the primary conclusions and associated recommendations given there. Rather, the new findings strengthen those conclusions and recommendations.
2. Overall, the evidence and logical basis for recommendations concerning prevention of occupational respiratory disease among coal

miners remains essentially unaffected by the newer findings that have emerged since publication of the CCD.

In summary, as recommended by the CCD, every effort needs to be made to reduce exposure to both coal mine dust and to crystalline silica dust. As also recommended in the CCD, the latter task requires establishing a separate compliance standard in order to provide an effective limit to exposure to crystalline silica dust.

ABBREVIATIONS

CAO	chronic airway obstruction
CCD	coal criteria document, formally *NIOSH Criteria for a Recommended Standard—Occupational Exposure to Respirable Coal Mine Dust*
COPD	chronic obstructive pulmonary disease
CWP	coal workers' pneumoconiosis
CWXSP	Coal Workers' X-ray Surveillance Program
FEV_1	forced expiratory volume in 1 second
ILO	International Labour Office
mg/m^3	milligrams per cubic meter
MSHA	Mine Safety and Health Administration
NIOSH	National Institute for Occupational Safety and Health
PAH	polycyclic aromatic hydrocarbons
PDM	personal continuous dust monitor
PMF	progressive massive fibrosis
REL	recommended exposure limit
YPLL	years of potential life lost

GLOSSARY

Aerodynamic diameter: The diameter of a sphere with a density of 1 g/cm^3 and with the same stopping time as the particle. Particles of a given aerodynamic diameter move within the air spaces of the respiratory system identically, regardless of density or shape.

Chronic obstructive pulmonary disease (COPD): Includes chronic bronchitis (inflammation of the lung airways associated with cough and phlegm production), impaired lung function, and emphysema (destruction of the air spaces where gas transfer occurs). COPD is characterized by irreversible (although sometimes variable) obstruction of lung airways, and should be considered in any patient who has dyspnea, chronic cough or sputum, and/or a history of exposure to risk factors for COPD. The diagnosis should be confirmed by spirometry.

Coal rank: A classification of coal based on fixed carbon, volatile matter, and heating value of the coal. Coal rank indicates the progressive geological alteration (coalification) from lignite to anthracite.

Coal workers' pneumoconiosis (CWP): A chronic dust disease of the lung arising from employment in a coal mine. In workers who are or have been exposed to coal mine dust, diagnosis is based on the radiographic classification of the size, shape, profusion, and extent of parenchymal opacities.

Crystalline silica: Silicon dioxide (SiO_2). "Crystalline" refers to the orientation of SiO_2 molecules in a fixed pattern as opposed to a nonperiodic, random molecular arrangement defined as amorphous. The three most common crystalline forms of free silica encountered in general industry are quartz, tridymite, and cristobalite. The predominant form is quartz.

Excess (exposure-attributable) prevalence: The prevalence (cases/ population at risk) attributable to workplace dust exposure (in the case of CWP, the prevalence adjusted for radiographic appearances associated with lung aging).

International Labour Office (ILO) classification system: A standardized method for assessing abnormalities related to the pneumoconioses based substantially on comparison of test with reference radiographs. In the system there are 4 categories of simple pneumoconiosis (categories 0, 1, 2, and 3), with 0 implying no definite abnormality.

Progressive massive fibrosis: Coal workers' complicated pneumoconiosis. Diagnosis is based on determination of the presence of large opacities (1 cm or larger) using radiography or the finding of specific lung pathology on biopsy or autopsy.

Quartz: Crystalline silicon dioxide (SiO_2) not chemically combined with other substances and having a distinctive physical structure.

Respirable coal mine dust: That portion of airborne dust in coal mines that is capable of entering the gas-exchange regions of the lungs if inhaled: by convention, a particle-size-selective fraction of the total airborne dust; includes particles with aerodynamic diameters less than approximately 10 μm.

ACKNOWLEDGMENTS

Michael Attfield was the primary author of this document, which was prepared in the Division of Respiratory Disease Studies, NIOSH, under the direction of Dr. David Weissman. He acknowledges the following assistance in the preparation of the report:

Division of Respiratory Disease Studies (DRDS)

Janet Hale, Eva Suarthana, Mei Lin Wang

Health Effects Laboratory Division (HELD)

Vincent Castranova, Kimberly Clough Thomas

This Current Intelligence Bulletin has undergone substantial internal and external scientific peer review with subsequent revision. A draft version of the document was published on the NIOSH website for public comment for 60 days, with notification of its availability via the Federal Register. All input, both internal and external, has been considered, addressed, and responded to in preparation of this final version.

Internal NIOSH reviewers who provided critical feedback to the preparation of the document

Eileen Kuempel, Education and Information Division (EID). Ed Thimons, Office of Mine Safety and Health Research (OMSHR), Robert Castellan (DRDS), Ainsley Weston (DRDS), Eileen Storey (DRDS).

External Expert Peer Review Panel

NIOSH expresses appreciation to the following independent, external reviewers for providing insights and comments that contributed to the development of this document and enhanced the final version:

Robert Cohen, M.D., F.C.C.P.
Cook County Health and Hospitals System
Chicago, IL

Dennis O'Dell
United Mine Workers of America
Washington, DC

Joseph Lamonica
Bituminous Coal Operators Association
Washington, DC

Cecile Rose, M.D.
National Jewish Health
Denver, CO

1. INTRODUCTION

The publication of the *NIOSH Criteria for a Recommended Standard—Occupational Exposure to Respirable Coal Mine Dust* or Coal Criteria Document (CCD) in 1995 (1) followed a long period of extensive research activity focused on exposure to coal mine dust and its health effects in coal miners. From this research, substantial information had emerged about the extent and severity of respiratory disease caused by coal mine dust, its quantitative relationship with dust exposure, its pathology and toxicology, environmental patterns of relevant exposures, and methodologies for assessing these variables. In particular, the findings demonstrated that not only was there a considerable burden of coal workers' pneumoconiosis (CWP) in the U.S. and other countries, but that underground coal miners were vulnerable to other lung diseases, notably chronic obstructive pulmonary disease (COPD). The evidence came from extensive and well-planned epidemiologic and laboratory-based investigations undertaken primarily in the U.S., the United Kingdom, and (West) Germany, with supporting information coming from studies in other European countries and Australia.

The available information was thoroughly summarized in the CCD. It showed that, in 1995, CWP was in decline in the U.S., with downward trends in prevalence in all tenure groups (Figure 4–2 of the CCD (1), included here as Figure 1). This decline was consistent with reductions in coal mine dust exposure mandated by the 1969 Coal Mine Health and Safety Act (1969 Coal Mine Act) (CCD Figure 4–1 (1); Figure 2). Despite this decline in disease levels, NIOSH concluded from review of the surveillance data and quantitative risk estimates based on the epidemiologic studies that the current dust exposure regulations for U.S. coal mines were not sufficiently protective. Consequently, it proposed lower dust limits for coal mines, enhanced medical surveillance, and other requirements.

The CCD noted that the current U.S. federal dust limit for underground coal mines dated back to the 1969 Coal Mine Act, which mandated a compliance permissible exposure limit of 2 milligrams per cubic meter (mg/m^3) of respirable coal mine dust. This limit was derived from British research, which provided the only quantitative exposure-response relationship available at that time. This exposure-response curve (CCD Figure 7–2 (1); Figure 3) predicted that no cases of CWP as severe as category 2 on the International Labour Office (ILO) classification system (3) would develop among miners who worked 35 years at 2 mg/m3. Similarly at that time, the current information indicated that the disabling form of CWP, progressive massive fibrosis (PMF), was very unlikely to develop from less severe ILO categories (e.g., category 1 CWP). Therefore, adoption of the 2 mg/m^3 limit was believed, at that time, to be protective against the risk of disability and premature mortality that accompanies PMF.

Subsequent scientific findings, emerging between 1969 and 1995, disproved some of the assumptions inherent in the adoption of the 2 mg/m^3 standard. Firstly, the assumption that miners with CWP less severe than category 2 were at minimal risk of PMF was found to be incorrect. Moreover, additional findings from the British data (CCD Figure 7–6 and Table 4–6 (1); Figure 4 and Table 1), together with new results on U.S. underground coal miners from research undertaken by NIOSH (CCD Figure 7–4 and Table 4–6 (1); Figure 5 and Table 1) showed that there was no threshold at 2 mg/m^3 as had been indicated by the original British study (CCD Figure 7–2 (1); Figure 3). Furthermore, the CCD reviewed findings on other lung diseases and their relationship with coal mine dust exposure. It concluded that coal miners were at additional risk of developing ventilatory function deficits, respiratory symptoms, and emphysema in addition to CWP (CCD Table 4–7 (1); Table 2). (Note that the results provided here are selected to illustrate the main CCD conclusions. For a full understanding of the complete body of knowledge, please see the CCD [1].)

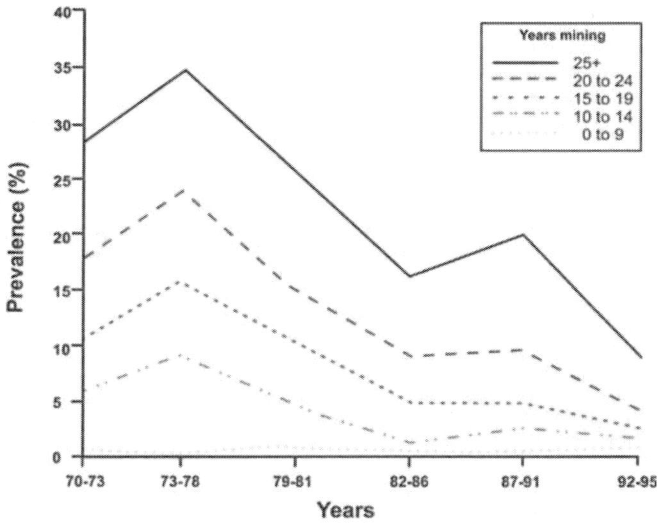

Figure 1. Prevalence of CWP category 1 or greater from the NIOSH Coal Workers' X-ray Program from 1970–1995, by tenure in coal mining. (Figure 4–2 of the CCD [1]).

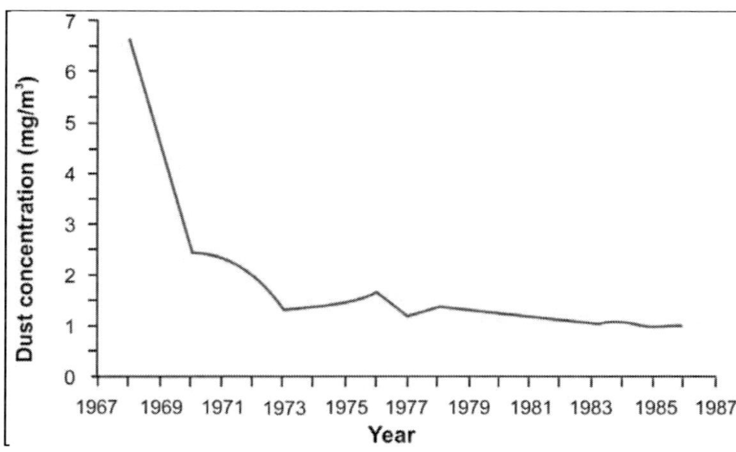

Source: Attfield and Wagner [2].

Figure 2. Trend in reported dust concentrations for continuous miner operators, 1968–87. [Figure 4–1 of the CCD (1)].

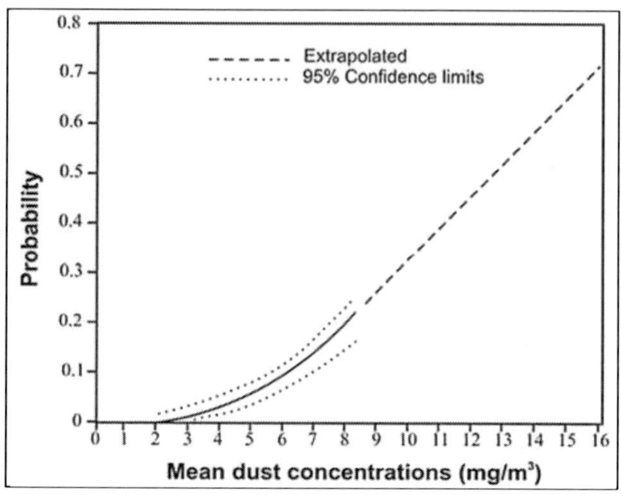

Source: Jacobsen et al. [4].

Figure 3. Probability that an individual starting with no pneumoconiosis (category 0/0) will be classified as 2/1 or greater after 35 years of exposure to various concentrations of coal mine dust. [Figure 7–2 of the CCD [1]].

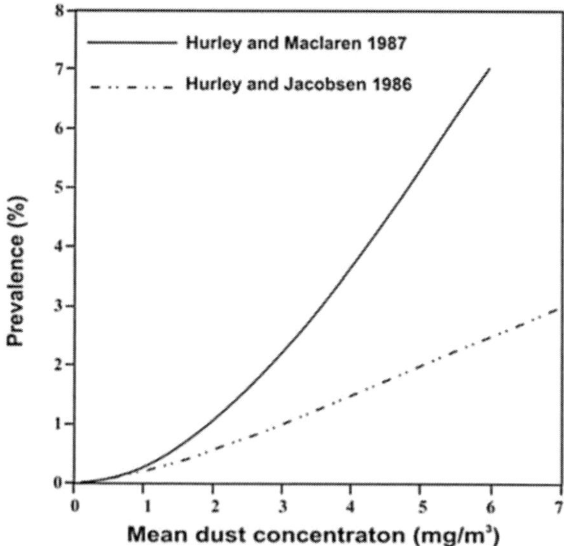

Source: Hurley and Maclaren [5].

Figure 4. Predicted prevalence of PMF among British coal miners after a 35-year working lifetime by mean concentration of respirable coal mine dust. [Figure 7–6 of the CCD [1]].

Table 1. Predicted prevalence of simple CWP and PMF among U.S. or British coal miners at age 58 following T exposure to respirable coal mine dust over a 40-year working lifetime. (Table 4–6 of the CCD[1])

Study and coal rank	Mean concentration of respirable CMD (mg/m^3)	Predicted prevalence (cases/1,000)		
		CWP≥1	CWP≥2	CWP≥3
Attfield and Seixas (6):*				
High-rank bituminous	2.0	253 (204–308)†	89 (60–130)	51 (30–85)
	1.0	116 (88–150)	29 (16–51)	16 (7–36)
Medium/low-rank bituminous	2.0	144 (117–176)	31 (20–49)	14 (7–27)
	1.0	84 (64–110)	17 (9–30)	9 (4–19)
Attfield and Morring (7):‡				
Anthracite	2.0	316 (278–356)	142 (118–172)	89 (69–113)
	1.0	128 (108–152)	46 (35–60)	34 (24–48)
High-rank bituminous (89% carbon)	2.0	282 (250–317)	115 (94–141)	65 (49–85)
	1.0	119 (100–142)	41 (31–54)	29 (20–41)
Medium/low rank bituminous (83% carbon)	2.0	121 (108–136)	40 (33–49)	22 (17–29)
	1.0	74 (62–89)	24 (18–31)	17 (12–24)
Medium/low-rank bituminous (Midwest)	2.0	89 (73–108)	28 (20–39)	15§ (9–26)
	1.0	63 (52–77)	20 (14–27)	14§ (9–21)
Medium/low rank bituminous (West)	2.0	67 (52–86)	15 (8–26)	13§ (7–24)
	1.0	55 (44–68)	14 (10–21)	12§ (8–20)
Hurley and Maclaren				
High-rank bituminous (89% carbon)	2.0	89	29	18
	1.0	40	12	7
Medium/low-rank bituminous (83% carbon)	2.0	65	16	7
	1.0	28	7	3

* Attfield and Seixas [6] define the coal rank groups as follows: 1- high-rank bituminous (89–90% carbon)—central Pennsylvania and southern West Virginia; 2-medium/low-rank bituminous (80–87% carbon)—western Pennsylvania, northern and southwestern West Virginia, eastern Ohio, eastern Kentucky, western Virginia and Alabama; 3- low-rank—western Kentucky, Illinois, Utah, and Colorado.

† Ninety-five percent confidence intervals, where available, are in parentheses under the point estimates for prevalence (cases/ 1000).

‡ In Attfield and Morring [7], the predicted prevalences for CWP category 1 or greater and category 2 or greater did not include PMF (correction from CCD original).

§ Attfield and Morring [7] define the coal rank groups as follows: 1- anthracite - two mines in eastern Pennsylvania (~93% carbon); 2- medium/low-volatile bituminous (89–90% carbon)—three mines in central Pennsylvania and three in southeastern West Virginia; 3- high-volatile "A" bituminous (80–87% carbon)—16 mines in western Pennsylvania, north and southwestern West Virginia, eastern Ohio, eastern Kentucky and Illinois.

Table 2. Predicted prevalence of decreased lung function* among U.S. or British coal miners at age 58 following exposure to respirable coal mine dust over a 40-year working lifetime. (Table 4–7 of the CCD [1])

Study and region†	Mean concentration of respirable coal mine dust (mg/m^3)	Lung function decrement (% FEV_1)	Predicted prevalence (cases/1,000)	
			Never smokers	Smokers
Attfield and Hodous [8]				
East	2.0	<80	141	369
		<65	22	102
	1.0	<80	123	336
		<65	18	87
West	2.0	<80	125	340
		<65	16	80
	1.0	<80	108	309
		<65	13	68
Marine et al. [9]:				
	2.0	<80	153	372
		<65	63	173
	1.0	<80	125	314
		<65	52	159

* Decreasing lung function is defined as FEV_1 <80% of predicted normal values. Clinically important deficits are FEV_1 <80%, which approximately equals the lower limit of normal (LLN), or the fifth percentile [9, 10]; and FEV_1 <65%, which has been associated with severe exertional dyspnea [11, 12].

† Attfield and Hodous [13] define the following coal rank regions: East—anthracite (eastern Pennsylvania) and bituminous (central Pennsylvania, northern Appalachia [Ohio, northern West Virginia, western Pennsylvania], southern Appalachia [southern West Virginia, eastern Kentucky, western Virginia]), Midwest (Illinois, western Kentucky), and South (Alabama).

‡ Conversion from gh/m^3 to mg-yr/m^3; assumed 1,920 hr/yr for U.S. miners.

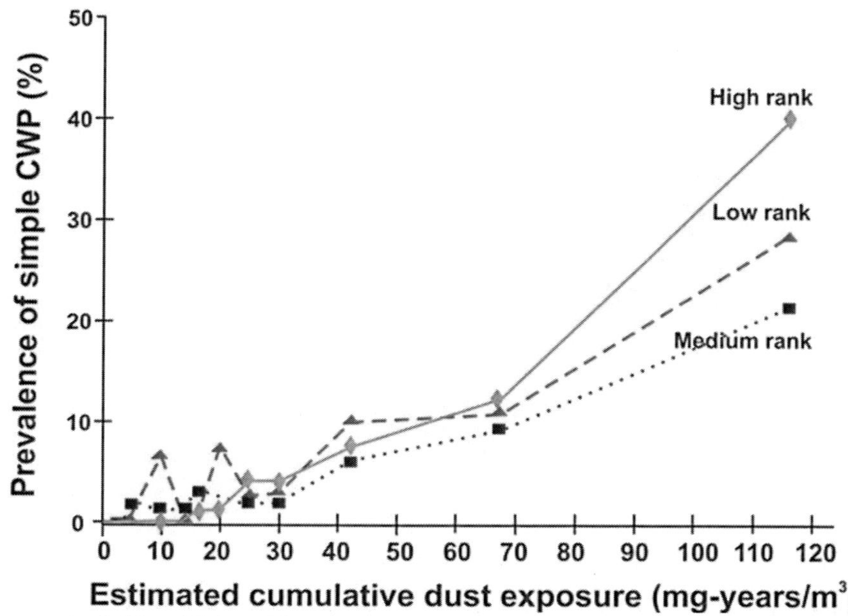

Source: Attfield and Seixas [6].

Figure 5. Prevalence of simple CWP category 1 or greater among U.S. coal miners by estimated cumulative coal mine dust exposure and coal rank. [Figure 7–4 of the CCD [1]].

On the basis of the updated body of evidence on the adverse health effects and an evaluation of the technological feasibility of reducing dust exposures, NIOSH recommended that the federal coal mine dust limit be reduced to 1 mg/m^3. Critical to this decision were computed excess (exposure-attributable) prevalences of CWP derived from two studies of U.S. coal miners undertaken by NIOSH (CCD Table 7–2 (1); Table 3). Predictions were derived from each study for a working lifetime (i.e., 45 years) exposure at 2 mg/m^3 and 1 mg/m^3. Another source of information critical to the recommendations was information on predicted excess lung function decrements for a lifetime's exposure to 1 mg/m^3 and 2 mg/m^3 respectively (CCD Table 7–3 (1); Table 4). Details of the rationale for, and development of, the risk analyses employed in the CCD were subsequently published separately (8). NIOSH also evaluated information from other epidemiologic studies in coming to its recommendations in the CCD. However, because no other studies had quantitative exposure-response information, apart from strengthening the case for more stringent regulation, these additional study results did not provide any numerical basis for standard setting. For simplicity, NIOSH recommended one exposure limit for the nation rather than different limits by coal rank, based on technological feasibility of reducing exposures, even though CWP prevalence has been shown to vary according to the rank of the coal in studies of miners in the U.S. and other countries.

Table 3. Excess (exposure-attributable) prevalence of simple CWP or PMF among U.S. coal miners at age 65 following exposure to respirable coal mine dust over a 45-year working lifetime. (Table 7–2 of the CCD [1])

Study and coal rank	Disease category	Cases/1,000 at various mean dust concentrations		
		0.5 mg/m^3	1.0 mg/m^3	2.0 mg/m^3
Attfield and Seixas [6]:*				
High-rank bituminous	CWP ≥ 1	48	119	341
	CWP ≥ 2	20	58	230
	PMF	13	36	155
Medium/low-rank bituminous	CWP ≥ 1	27	63	165
	CWP ≥ 2	9	22	65
	PMF	4	10	29
Attfield and Morring [7]:†				
Anthracite	CWP ≥ 1	45	120	380
	CWP ≥ 2	17	51	212
	PMF	17	46	167
High-rank bituminous (89% carbon)	CWP ≥ 1	41	108	338
	CWP ≥ 2	15	43	168
	PMF	13	34	114
Medium/low-rank bituminous (83% carbon)	CWP ≥ 1	18	42	111
	CWP ≥ 2	6	15	42
	PMF	4	9	21
Medium/low-rank bituminous (Midwest)	CWP ≥ 1	12	26	64
	CWP ≥ 2	4	9	22
	PMF	1	3	6
Medium/low-rank bituminous (West)	CWP ≥ 1	7	14	32
	CWP ≥ 2	<1	<1	1
	PMF	<1	<1	1

* Attfield and Seixas [6] define the coal rank groups as follows: 1- high-rank bituminous (89–90% carbon)—central Pennsylvania and southern West Virginia; 2- medium/low-rank bituminous (80–87% carbon)—western Pennsylvania, northern and southwestern West Virginia, eastern Ohio, eastern Kentucky, western Virginia and Alabama; 3- low-rank—western Kentucky, Illinois, Utah and Colorado.

† Attfield and Morring [7] define the coal rank groups as follows: 1- anthracite - two mines in eastern Pennsylvania (~93% carbon); 2- medium/low-volatile bituminous (89–90% carbon)—three mines in central Pennsylvania and three in southeastern West Virginia; 3- high-volatile "A" bituminous (80–87% carbon)—16 mines in western Pennsylvania, north and southwestern West Virginia, eastern Ohio, eastern Kentucky and Illinois.

Table 4. Excess (exposure-attributable) prevalence of decreased lung function[*] among U.S. coal miners at age 65 following exposure to respirable coal mine dust over a 45-year working lifetime. (Table 7–3 of the CCD [1])

Study and region	Lung function decrement	Smoking status	Cases/1,000 at various mean dust concentrations		
			0.5 mg/m^3	1.0 mg/m^3	2.0 mg/m^3
Attfield and Hodus [8]:[†]					
East	<80% FEV_1	Never smoked	10	21	44
		Smoker	12	24	51
West	<80% FEV_1	Never smoked	9	19	40
		Smoker	11	23	48
East	<65% FEV_1	Never smoked	2	5	12
		Smoker	4	8	19
West	<65% FEV_1	Never smoked	2	4	9
		Smoker	3	7	15
Seixas et al. [14]:	<80% FEV_1	Never smoked	60	134	315
		Smoker	68	149	338
	<65% FEV_1	Never smoked	18	45	139
		Smoker	27	67	188

[*] Decreasing lung function is defined as FEV_1 <80% of predicted normal values. Clinically important deficits are FEV_1 <80%, which approximately equals the lower limit of normal (LLN), or the fifth percentile (9, 10); and FEV_1 <65%, which has been associated with severe exertional dyspnea [11, 12].

[†] Attfield and Hodous [8] define the following coal rank regions: East—anthracite (eastern Pennsylvania) and bituminous (central Pennsylvania, northern Appalachia [Ohio, northern West Virginia, western Pennsylvania], southern Appalachia [southern West Virginia, eastern Kentucky, western Virginia]), Midwest (Illinois, western Kentucky), and South (Alabama).

[‡] Coal rank was not provided in Seixas et al. [14] However, miners were included from bituminous coal ranks and regions across the United States, as described in Attfield and Seixas [6]: 1. High-rank bituminous (89%–90% carbon): central Pennsylvania and southeaster West Virginia; 2. Medium/low-rank bituminous (80%–87% carbon): medium-rank—western Pennsylvania, northern and southwestern West Virginia, eastern Ohio, eastern Kentucky, western Virginia, and Alabama; low-rank—western Kentucky, Illinois, Utah, and Colorado.

In addition to recommending a reduction in the exposure limit for coal mine dust, NIOSH also recommended a change in the exposure limit for crystalline silica dust and the method by which it is enforced. Currently, silica levels are intended to be controlled by a reduction in the level of coal mine dust commensurate with the proportion of the dust that is silica. NIOSH proposed a separate limit for respirable crystalline silica in order to more effectively monitor and control exposures. The NIOSH recommendations for coal mine dust and crystalline silica dust were explicitly intended for both underground and surface coal operations. In addition, NIOSH recommended enhancing worker medical monitoring, and extending it to surface coal mine workers. An independent advisory committee, which was convened by MSHA in 1996 in response to the NIOSH CCD, affirmed each of the recommendations in the CCD [15].

The following material comprises a summary of results from reports that have been published since 1995 on CWP, other respiratory diseases, cancer outcomes, and overall mortality. In addition, new information summarized from the current NIOSH medical monitoring program for coal miners is included. There are also sections on aspects relating to dust levels, control, and compliance, and on surface coal mining.

2. Coal Workers' Pneumoconiosis

2.1. Surveillance

Over time since 1995 it has become increasingly apparent that the observed prevalence of CWP in U.S. underground coal miners examined in the Coal Miners' X-ray Surveillance Program (CWXSP) was no longer declining as it had from 1969–1995, but had begun increasing. This situation was first noticed in a 2003 CDC/NIOSH report [16]. This report also drew attention to the fact that CWP was developing in underground coal miners who had spent all of their working life in a working environment where the dust conditions should have been as mandated by the 1969 Coal Mine Act. Based on findings that showed higher CWP prevalences in certain worker groups, the publication raised concerns about possible excessive dust exposures in certain states, at smaller mines, and by some surface and contract miners.

Reports in 2006 and 2007 called attention to advanced pneumoconiosis in working underground miners in Kentucky (KY) and Virginia (VA) [17, 18]; as with the prior report (2003), most of the affected miners had started work after 1969 yet had still developed severe CWP. Possible reasons put forth as explanations for the findings were: 1) inadequacies in the mandated coal mine dust regulations; 2) failure to comply with or adequately enforce those regulations; 3) lack of disease prevention innovations to accommodate changes in mining practices (e.g., thin-seam mining); and 4) missed opportunities by miners to be screened for early disease and take action to reduce dust exposures. Further explanations, noted in other reports included: 5) longer hours being worked by today's coal miners; 6) excessive exposure to crystalline silica, perhaps associated with the mining of thinner seams of coal; and 7) lack of resources for dust control and miner/operator education, particularly in smaller mines [19–21].

To gain a better understanding of the extent of the problem, NIOSH undertook a systematic analysis of rapidly progressive CWP [22]. Statistics were derived based on each miner's radiographic steps of progression of CWP using the standard ILO categorical scores standardized to a five-year interval. These data were summarized by county and then plotted to reveal 'hot spots' of rapid disease progression (Figure 6; from Antao et al, 2005 [22]). These tended to be located on the eastern edge of the Appalachian coal field but were particularly concentrated in the southern West Virginia (WV)/western VA/eastern KY tri-state region (central Appalachian region).

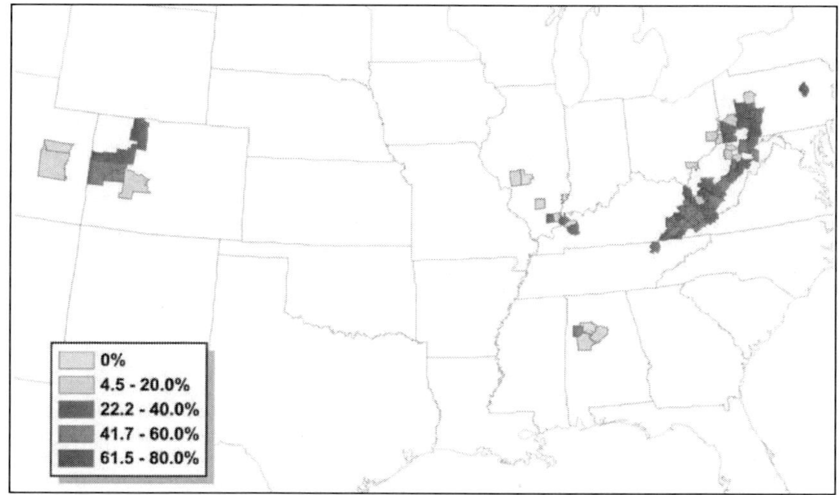

Source: Antao et al. [22].

Figure 6. Percentage of evaluated miners with rapidly progressive coal workers' pneumoconiosis by county (not shown are counties with fewer than five miners evaluated).

In response to these observations, NIOSH undertook a series of field surveys in the hot spot regions in an attempt to enhance the quality of data (i.e., improve participation). The field surveys were undertaken as part of the Coal Workers' X-ray Surveillance Program (CWXSP) administered by NIOSH, as mandated by the 1969 Coal Mine Act. The targeted surveys comprised an "Enhanced Program" to complement the regular CWXSP program. Those findings are included in an overall tabulation that can be found in the NIOSH *2007 Work-related Lung Disease (WoRLD) Surveillance Report* (disseminated in hard copy [23] and on the NIOSH internet site [24]. These results showed that the prevalence of CWP appeared to have stopped declining around 1995–1999, and has risen since then. The trend reversal appears most apparent in the longer-tenured miners. Virtually that entire group had spent their whole working life in dust conditions mandated by the 1969 Coal Mine Act. An updated (unpublished) version of this graphic, taking the data up to 2009, shows the increased CWP prevalence observed over the past decade (Figure 7).

The upward trend visible in Figure 7 for all pneumoconiosis cases (category 1+) is even more evident for PMF (Figure 8). Of particular concern are the prevalence values for the last three five-year periods (1995–2009) for miners with <25 years tenure, which are well above those observed in the early 1990s. In 2005–2009 alone, 69 coal miners examined in the CWXSP were determined to have PMF. Of these, 11 had less than 25 years total tenure in coal mining, and the majority (56, or 81%) were working in the central Appalachian region.

Since the data from 2005 in Figures 7 and 8 were derived, in part, from the special NIOSH surveys targeted at hot spot areas, there could be the concern that the recent CWXSP findings may be upwardly biased, with the implication that the apparent rise in prevalence may be an artifact. However, overall prevalences for 2005–2009 for the Enhanced and regular programs, derived from state-specific prevalences weighted by the participation rates for the whole program from the different states gives rise to figures of 3.2% prevalence for data from the targeted surveys (Enhanced program) compared to 3.1% for data from the regular program. There is therefore no indication whatsoever of any major bias. Moreover, it is clear from Figures 7 and 8 that the upswing in prevalence of CWP was underway before the targeted surveys began in 2005.

The finding of severe CWP in the CWXSP was confirmed among West Virginia coal miners with a report of 138 compensated cases of PMF from 2000–2009 (25). These miners had worked virtually all of their lives under post-1969 Coal Mine Act conditions, and had developed PMF at age 52.6 years on average. Of the 138, 21 had died by publication date.

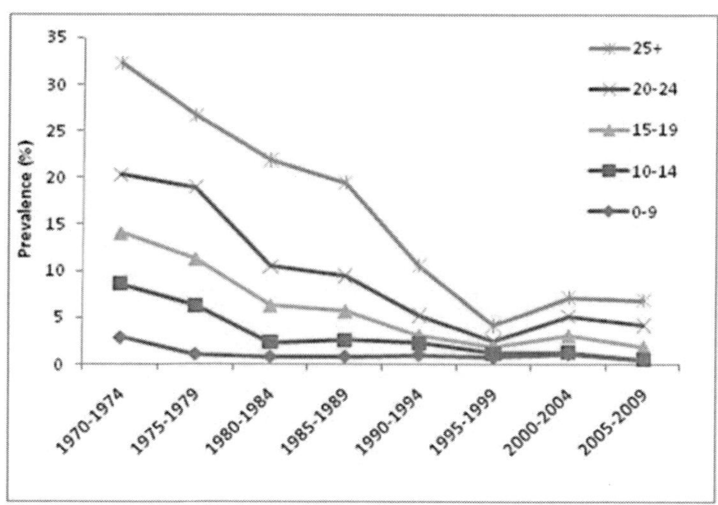

Source: NIOSH CWXSP data.

Figure 7. Percentage of miners examined with CWP category 1 or greater from the NIOSH Coal Workers' X-ray Program from 1970–2009, by tenure in coal mining.

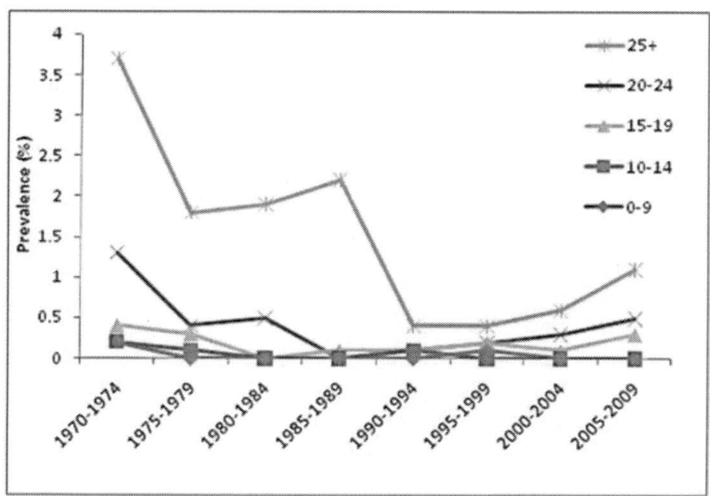

Source: NIOSH CWXSP data.

Figure 8. Percentage of miners examined with PMF from the NIOSH Coal Workers' X-ray Program from 1970–2009, by tenure in coal mining.

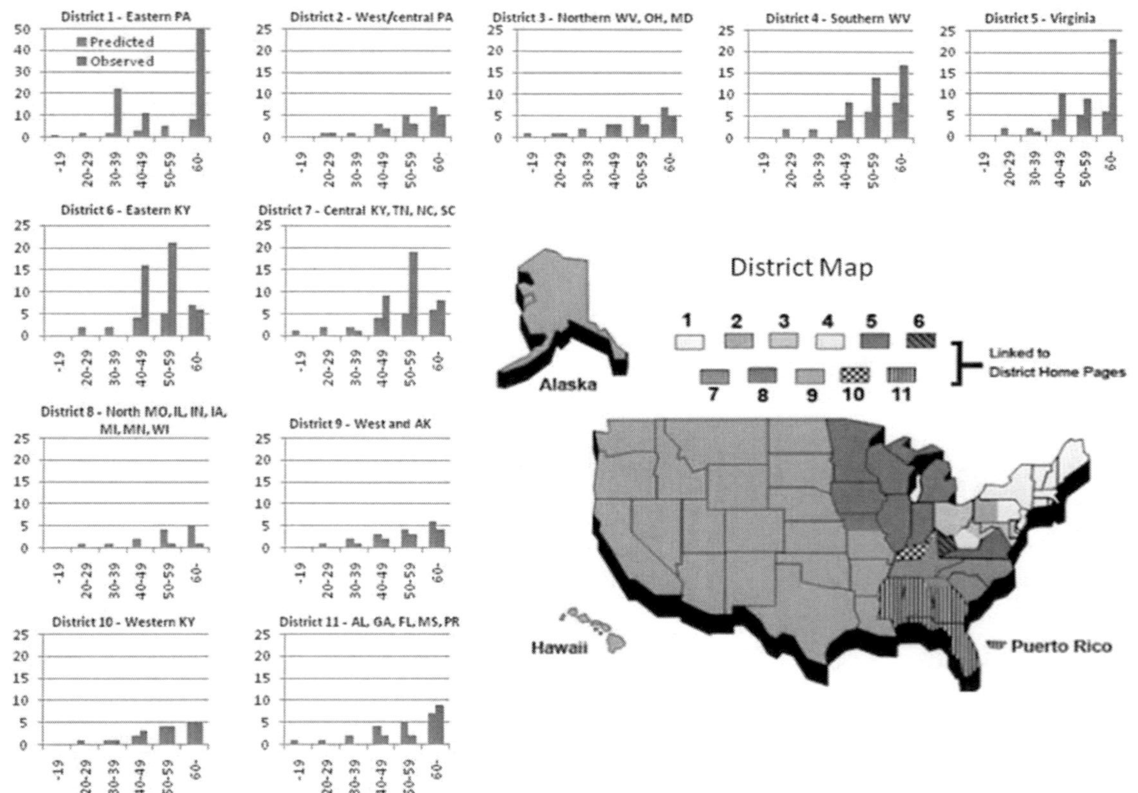

Source: Data from the NIOSH CWXSP for 1995–2009. Charts based on data provided in Suarthana et al [32].

Figure 9. Observed and predicted prevalences (%) of CWP category 1 or greater by age group and MSHA District.

A number of reports of surveillance information from other countries have emerged since 1995 [26–31]. Although mining conditions differ in these other countries, these studies are supportive of the findings in U.S. coal miners.

2.2. Epidemiology

The CWXSP finding of increased prevalence of CWP has prompted a series of analyses by NIOSH aimed at identifying factors that might be causing the increase. In an analysis undertaken as part of this series [32], predicted risk of CWP was derived for each individual who participated in the CWXSP from 2005–2009 using published exposure-response models [7]. The models were coal-rank specific, and used both age and cumulative coal mine dust exposure as predictors. The resulting individual risks (lying between 0 and 1) were then summed over subsets of the data and compared with the observed prevalences. (Further models, published later, were also available, but the early relationships were preferred because they were based on greater numbers of observations and had more specific adjustment for coal rank. The later predicted prevalences were somewhat higher than those presented here.) The results tabulated by MSHA region and age are shown here in Figure 9. It is clear that CWP prevalence is less than expected in some regions (observed ≤ predicted) but substantially greater than expected in others (observed > predicted). That is, in the northern Appalachian region and the mid-west and western coal fields the observed prevalences are generally below those predicted in all age groups. However, in the southern WV, eastern and central KY, Tennessee, and VA MSHA regions the observed prevalences are 2–4 times greater than predicted from cumulative coal mine dust exposure and age. Clearly, some factor or factors must be acting differently across the regions to cause this regional pattern.

At least three environmental factors impact the central Appalachian region in this respect. These are: thin seams, small mines, and, for VA, high coal rank. The mining of thin coal seams, which often involves the deliberate cutting and extraction of substantial amounts of (often siliceous) rock overlying or underlying the coal seam, is particularly prevalent in Appalachia [33]. NIOSH has been investigating the health implications of possible excessive crystalline silica exposure arising from the cutting of the rock adjacent to the coal seams [34]. This analysis used the presence of r-type pneumoconiotic opacities on the chest X-ray as an indicator of crystalline silica exposure. This type of opacity is a radiographic manifestation of nodules in the lung having a typology often associated with excessive exposure to silica dust. An increase in the prevalence of such opacities could then well indicate that miners are more frequently being exposed to crystalline silica dust, or are experiencing exposure to higher levels of silica dust. Increased exposure to crystalline silica dust may well be arising from industry trends, whereby there is greater focus on mining thinner seams of coal as the more productive thicker seams are mined out.

The findings from this study indicated that the proportion of radiographs showing r-type opacities increased during the 1990s, and particularly after 1999, in KY, VA, and WV, compared to the 1980s. They could potentially be explained by an increase in the frequency and/or intensity of silica exposure among underground coal miners. This hypothesis was confirmed by evidence from dust sampling in mines in that region indicating that excessive silica exposures are occurring (35). In the CCD, NIOSH not only recommended that compliance procedures for crystalline silica be made more effective, but that the exposure limit be reduced. In the light of these epidemiological findings (34), therefore, this recommendation remains appropriate and even more urgent. A report on British coal miners also associated an increase in CWP prevalence with rock cutting (36).

In another NIOSH analysis (19) trends in CWP prevalence were examined by mine size (i.e., employment). The hypothesis investigated was that smaller mines lack resources in many areas for the full protection of workers, including dust suppression and up-to-date knowledge of means to prevent disease development. (It may also be that smaller mines tend to be those mining the thinner coal seams.) The results show that CWP prevalence is increasing in mines of all sizes, but the trend is much more obvious and much greater among miners employed in smaller coal mines.

Of the other factors listed above that may be contributing to the rise in CWP prevalence, increased working hours gives rise to special concern. Overall, U.S. coal miners are working longer hours. Figure 10, derived from data collected by MSHA shows a steady increase in the number of hours worked. This increase appears to arise from not only working longer shifts (e.g., 10 or 12 hours), but from working on weekends as well. Although no epidemiologic data exist that implicate longer hours as a contributory causative factor for CWP, working longer hours leads to the inhalation of more dust into the lungs. For example, working 12 hours leads to 50% more dust entering the lung compared to a regular 8-hour shift, assuming all other factors are equal (e.g., exposure concentration and breathing rates). Additionally, working longer workshifts reduces the time available between workshifts for the process of clearing the dust deposited in the lungs. Unfortunately, the available information on working hours in U.S. coal miners is not miner-specific but rather by coal mine, substantially reducing the validity of a formal analysis of this hypothesis. A report on British miners concluded that longer working hours were a factor in causing an increase in CWP prevalence at two mines [36]. In the CCD, NIOSH recommended reducing dust exposures below the 1 mg/m^3 REL for work shifts exceeding 40 hr/week (using the method of Brief and Scala [37]). This approach has also been recommended for British coal mines [38].

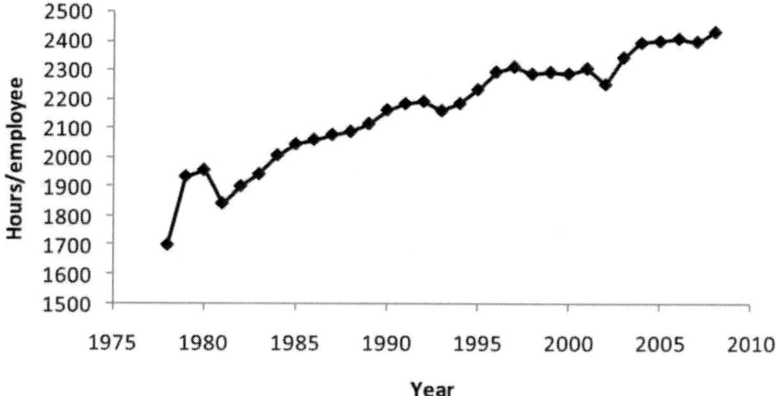

Source: http://www.msha.gov/ACCINJ/BOTHCL.HTM

Figure 10. Hours worked per underground coal miner, 1978–2008.

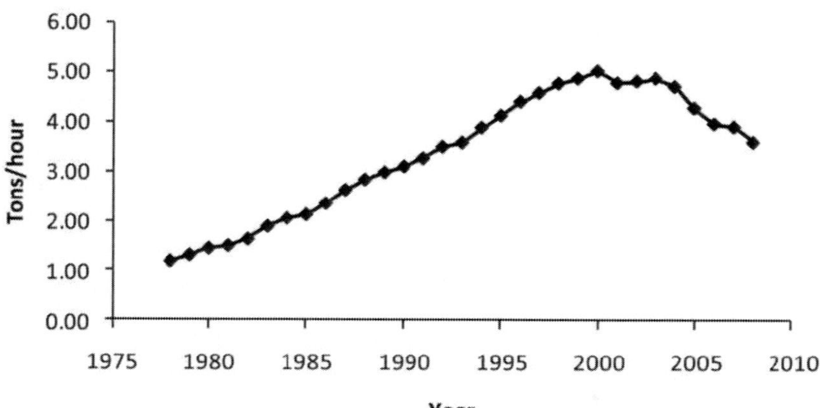

Source: http://www.msha.gov/ACCINJ/BOTHCL.HTM

Figure 11. Tons produced per hour worked at underground coal mines. 1978–2008.

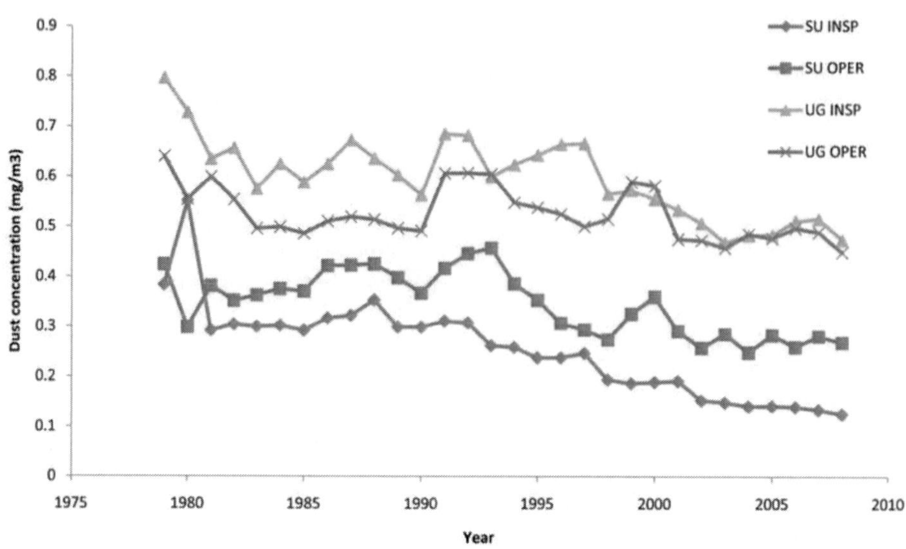

Figure 12. Respirable coal mine dust: Geometric mean exposures by type of mine (UG=underground, SU=surface), MSHA inspector (INSP) and mine operator (OPER) samples. [MSHA coal mine inspector and mine operator dust data].

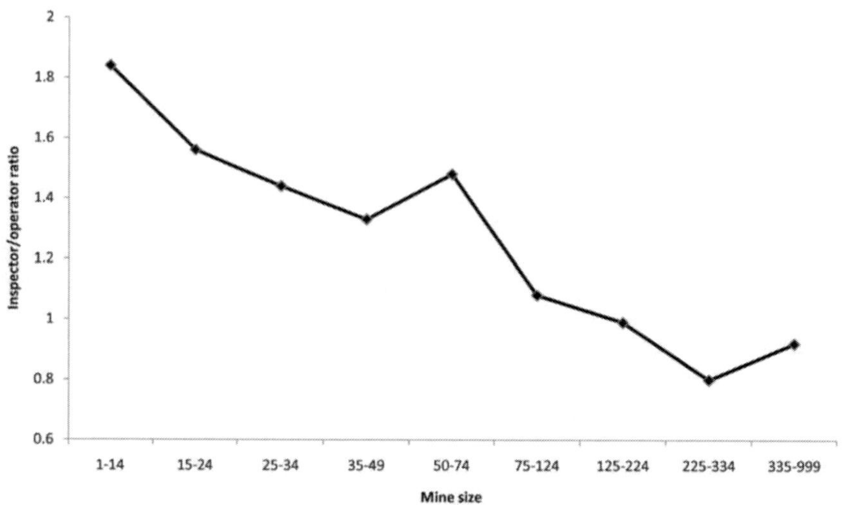

Source: MSHA Report of the Statistical Task Team of the Coal Mine Respirable Dust Task Group. [40]

Figure 13. Ratio of special inspection sample values to preceding operator compliance sample values by mine size.

Finally, productivity per hour worked also increased from 1978–2000, although it has since declined (Figure 11). Of course, these increases in productivity (and, presumably, increased potential for dust generation) should have been met by commensurate increases in dust prevention measures (e.g., ventilation and water sprays) in order to maintain compliance with the permissible exposure limit. Superficially, the current data appear to confirm this, in that airborne dust levels have apparently not risen during that period (Figure 12). However, the veracity of coal mine dust data has been challenged in the past (39). Moreover, the discovery of abnormal white centers in the dust sampling filters prompted a special inspection program [40] that showed that dust levels from operator samples consistently were lower than those from MSHA inspector samples, and that these differences were greater the smaller the mine (Figure 13). As with hours worked, there is a lack of reliable productivity data linkable with the health outcome data in order to investigate this issue further.

2.3. Mortality

A report on temporal patterns in pneumoconiosis mortality in the U.S. showed a substantial decline in numbers of deaths from CWP between 1968 and 2000 [41]. This decline is consistent with the reductions in dust level mandated by the 1969 Coal Mine Act. A major additional factor contributing to the declining number of CWP deaths is the diminishing coal mining workforce in the U.S. Figure 14 (Figure 1, from CDC [42]) shows the CWP death rate results extended to 2006. A similar situation has been observed in other developed countries, e.g., Australia [26]. Recent U.S. results have shown, however, a disconcerting increase in years of potential life lost (YPLL) due to CWP in the U.S. since 2002 [42]. Not only has the YPLL been increasing in younger CWP decedents (<65 years old), but the YPLL per CWP decedent has also been increasing over those same years (Figure 15; Figure 2 from CDC [42]). This may be related to the observed increase in CWP prevalence observed in recent years as noted earlier.

The post-1995 period saw the publication of a number of mortality analyses that augmented the earlier mortality findings on coal miners.

Most of those studies outside of the United States and United Kingdom did not have quantitative measurements of dust exposure. However, they do support previous findings concerning the overall increased mortality of coal miners and the additional risk imposed by the development of CWP [43–45].

Studies using quantitative exposures showed that mortality from CWP increased with increasing cumulative exposure to coal mine dust [46]. The British study [47] included exposure estimates for respirable quartz; cumulative exposure to respirable coal mine dust and respirable quartz were each highly significant predictors of pneumoconiosis mortality, although the relationship was stronger with coal dust than quartz. Respirable quartz exposure was associated with a small but statistically significant relative risk for lung cancer mortality [47].

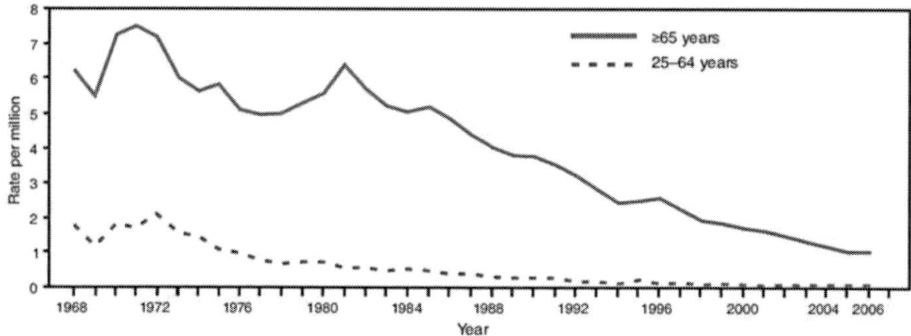

Source: CDC [42].

Figure 14. Age-adjusted death rates (per million) for decedents age ≥25 years with coal workers' pneumoconiosis as the underlying cause of death—United States, 1968–2006.

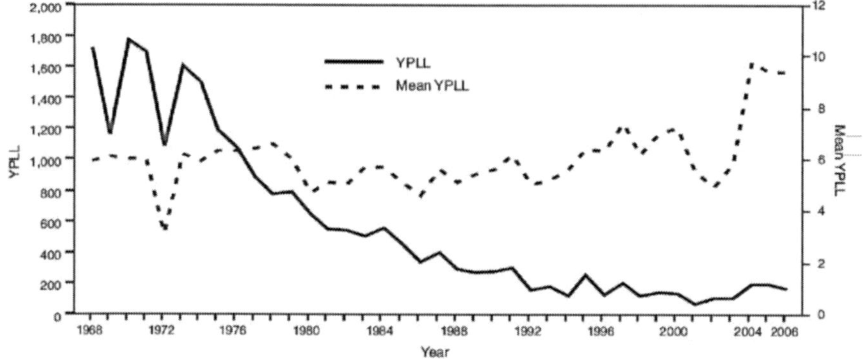

Source: CDC [42].

Figure 15. Years of potential life lost (YPLL) before age 65 and mean YPLL per decedent for decedents aged ≥25 years with coal workers' pneumoconiosis as the underlying cause of death—United States, 1968–2006.

2.4. Toxicology

Although coal mine dust and crystalline silica dust remain the two exposures of primary concern for environmental control, the post-1995 period has seen the publication of results from analyses aimed at eliciting information on what constituents of coal mine dust predict CWP development. These include: 1) free radicals, in which particles from freshly-fractured siliceous rock have been found to be more fibrogenic than aged particles [48]; 2) particle occlusion, in which clay present in the rock strata can surround the silica particles and render them less toxic [49]; and 3) bioavailable iron, which has been found to predict coal mine dust toxicity [50, 51].

McCunney et al. [52], favored the third explanation (bioavailable iron) and downplayed the role of quartz in the etiology of coal workers' pneumoconiosis. However, in an analysis of lung inflammatory cell counts from bronchoalveolar lavage in coal miners and non-miners, Kuempel et al. [53] showed that quartz dust (as either cumulative exposure or estimated lung burden) was a significant predictor of pulmonary inflammation and radiographic category of simple CWP. Cumulative coal dust exposure did not significantly add to those predictions, which may have been due to the high correlation between the coal and quartz cumulative exposures, such that separate effects for these two dusts could not be clearly demonstrated.

Against this, epidemiologic research has not demonstrated a strong effect of crystalline silica on CWP development in situations where silica levels are low. Rather, the level of coal mine dust, per se, has been the strongest predictor of CWP. However, the work of Laney et al. [34], as noted above, showed clear evidence of an increase in r-type radiographic opacities (typically associated with silicosis) and rapid progression of pneumoconiosis among U.S. coal miners in Kentucky, Virginia, and West Virginia, suggesting that they were exposed to excessive levels of respirable crystalline silica, and were thus at risk of silicosis. As noted previously, dust sampling results support this hypothesis [23]. There is, therefore, the clear need to minimize exposure to silica dust, especially for those jobs involving drilling or cutting sandstone and other siliceous rock. Moreover, as noted above, this is particularly pertinent because changing mining conditions might be leading to an increase in the potential for exposure to silica dust.

Page and Organiscak [54] linked the issue of coal rank, a known risk factor for CWP development in the U.S., Britain, and Germany, with the potential for higher levels of free radicals to be encountered where such coals are mined, and noted above by Dalal [48] and others to have greater levels of cytotoxicity.

2.5. Risk Analysis

Kuempel et al. [13] describe in more detail the risk analyses provided in the NIOSH CCD, including the excess (exposure-attributable) prevalence of CWP and PMF in underground coal miners exposed to various levels of coal mine dust for a working lifetime (as shown in the CCD and also presented here in Table 1). More recent risk estimates have been provided from research on British coal miners (Figure 16, from Figure 1 of Soutar et al. [55]). The latter apply to coal composed of 86.2% carbon (coal rank) and to underground coal miners who work 40 years at the designated coal mine dust level. Risks of PMF range from 0.8% at 1.5 mg/m^3 to about 5% at 6 mg/m^3, while risks of category 2 or greater CWP range from about 1.5% at 1.5 mg/m^3 to about 9% at 6 mg/m^3. Note that due to the different ways in which the risk estimates are derived, these are not directly comparable with those from U.S. studies shown in Table 1. However, the findings are consistent with those from U.S. studies in indicating that even at the lower coal mine dust levels recommended by NIOSH, and as noted in the CCD, some incidence of CWP would still be expected, especially among miners of higher rank coal.

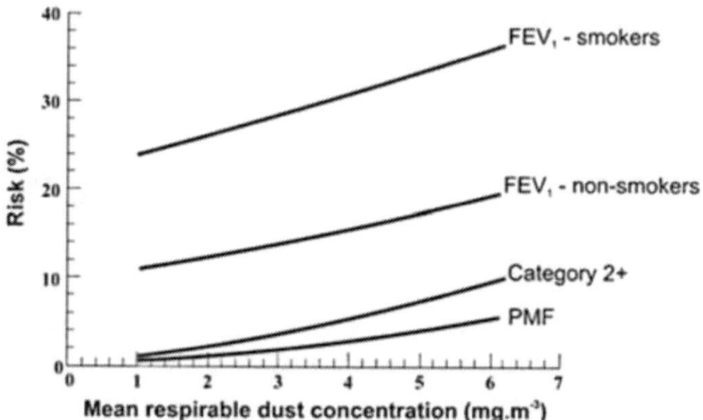

Source: Soutar et al. [55].

Figure 16. Risks at age 58–60 after 35–40 working years of: PMF; category 2 or greater (2+); 993 ml deficit of FEV_1 in nonsmokers; 993 ml deficit of FEV_1 in smokers.

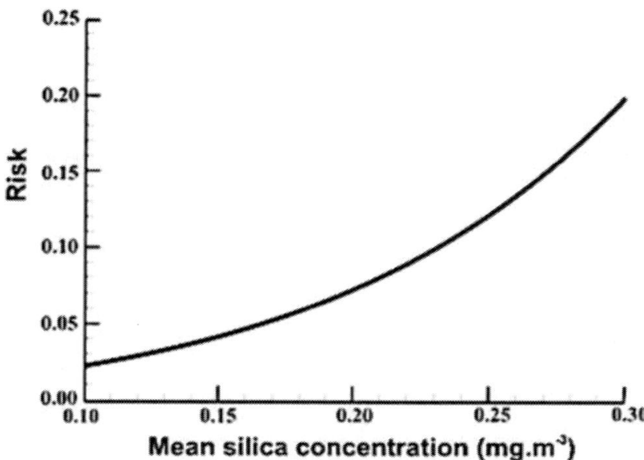

Source: Soutar et al. [55].

Figure 17. Risks for category 2 silicosis in relation to respirable silica concentration (<2 mg/m^3) averaged over 15 years.

Soutar et al. [55] also provide information on the risk of silicosis in underground coal miners. Their findings were developed from observations at one mine in which unusually high concentrations of crystalline silica dust occurred periodically (56). In their analysis, the authors chose to divide the analysis between exposures < 2 mg/m^3 and ≥ 2 mg/m^3. This dichotomy, in the authors' presentation, was associated with more rapid development of silicosis in the ≥ 2 mg/m^3 exposure range compared to chronic silicosis development at exposures < 2 mg/m^3. The findings indicate that short excursions to high silica dust intensities are considerably more hazardous than the same level of cumulative exposure at a lower intensity. They therefore demonstrate that mining situations involving the cutting of rock should be avoided if at all possible, or if necessary, that all precautions should be taken to minimize dust exposures. The findings for < 2 mg/m^3 (which apply to most coal mining environments that do not involve direct rock cutting) are given in Figure 17 (Figure 3 of Soutar et al. [55]).

3. Other Respiratory Disease Outcomes

Coggon and Taylor (57), in an extensive review, concluded that the "...balance of evidence points overwhelmingly to an impairment of lung function from exposure to coal mine dust, and this is consistent with the increased mortality from COPD that has been observed in coal miners." Findings on COPD and related outcomes in coal miners since 1995 [58–64] have continued to support their conclusion (which was largely based on pre-1995 information). The findings have also identified other risk factors in coal mining for pulmonary disease development. These include work in roof bolting, exposure to explosive blasting fumes, and exposure to dust control spray water previously stored in holding tanks [65].

The post-1995 findings have also elucidated patterns of lung function decline in coal miners, indicating that new miners tend to suffer more severe declines on starting work, after which the declines attenuate somewhat. This finding, derived initially from the analysis by Seixas et al. [66], was explored further by Henneberger and Attfield [58], who confirmed that the temporal pattern of lung function decline was different in newly-hired coal miners as compared to experienced miners. A possible reason for this could be a healthy worker survival effect. A study to explore this issue further, undertaken on new Chinese coal miners, confirmed that starting work in coal mining led to large initial drops in lung function, after which lung function declined at a lesser rate [67]. In a follow-up analysis, the researchers reported that the development of respiratory symptoms consistent with bronchitis contributed to the early declines in lung function [68].

A recently published mortality study from the United States [46] comprised a longer follow-up of a study on the same cohort of underground coal miners published in 1995 [69]. It showed that mortality from chronic airway obstruction (CAO) was elevated. Smoking, pneumoconiosis, coal rank region, and cumulative coal mine dust exposure were all predictors of mortality from CAO. Dust exposure effects were observed within the never-smoker subset of the cohort. The observed dust-related relative risks for CAO were similar to those for pneumoconiosis. The findings showed dust-related effects for chronic bronchitis and emphysema as well as CAO. A recent British study re-affirmed that mortality from COPD was related to coal mine dust exposure [47]. Finally, the implications of COPD (due to coal mine dust exposure as well as smoking) in causing increased mortality was explored by examining mortality risk in relation to rates of ventilatory function decline in coal miners [70]. Rates of ventilatory decline 2–3 times the normal age-related decline were associated with distinct increases in subsequent mortality.

Past pathologic studies have shown that emphysema severity in coal miners is related to dust exposure. Recent studies on South African and U.S. coal miners confirmed these findings [71, 72]. Important additional information on this topic, using quantitative estimates of both coal mine dust exposure and smoking amount, has been recently published by Kuempel et al. [73]. These authors found a highly significant relationship between cumulative exposure to respirable coal mine dust and emphysema severity at autopsy, controlling for effects of smoking, age, and other variables. The effect of dust exposure was similar in magnitude to that of smoking, and was seen in the never-smoking subgroup. In a further analysis, Kuempel et al. established that exposure to coal mine dust can produce clinically important levels of emphysema in coal miners [74].

The above findings support the CCD's recommendation to reduce the permissible coal mine dust exposure limit in underground coal mines to prevent the development of COPD, the associated severe declines in lung function, and the ensuing premature mortality.

There have been several reports of interstitial disease associated with exposure to coal mine dust, perhaps representing a manifestation of CWP, although little systematic research on this topic has been undertaken [75, 76].

3.1. Risk Analysis

Kuempel et al. [13] describe in more detail the risk analyses summarized in the NIOSH CCD, including the excess (exposure-attributable) prevalence of lung function deficits in underground coal miners exposed to various levels of coal mine dust for a working lifetime (CCD Table 4–7 (1); Table 2). More recent risk estimates from research on U.K. coal miners have been published (Figure 15; Figure 1 of Soutar et al. [55]). They apply to coal composed of 86.2% carbon (coal rank) and to underground coal miners who work 35 years at specified coal mine dust levels ranging from 1 to 6 mg/m^3. Risks of a deficit of approximately 1 liter in forced expiratory volume in 1 second (FEV_1) among never smokers range from 10% at zero dust exposure to about 19% at 6 mg/m^3. The concomitant risks for smokers range from about 22% to 36%, respectively. Note that due to the different ways in which the risk estimates have been derived, these are not directly comparable with those shown from U.S. studies shown in Table 2. However, they are consistent with findings from U.S. studies in that even at the 1 mg/m^3 coal mine dust exposure limit recommended by the CCD, some occupational effect on ventilatory function is expected.

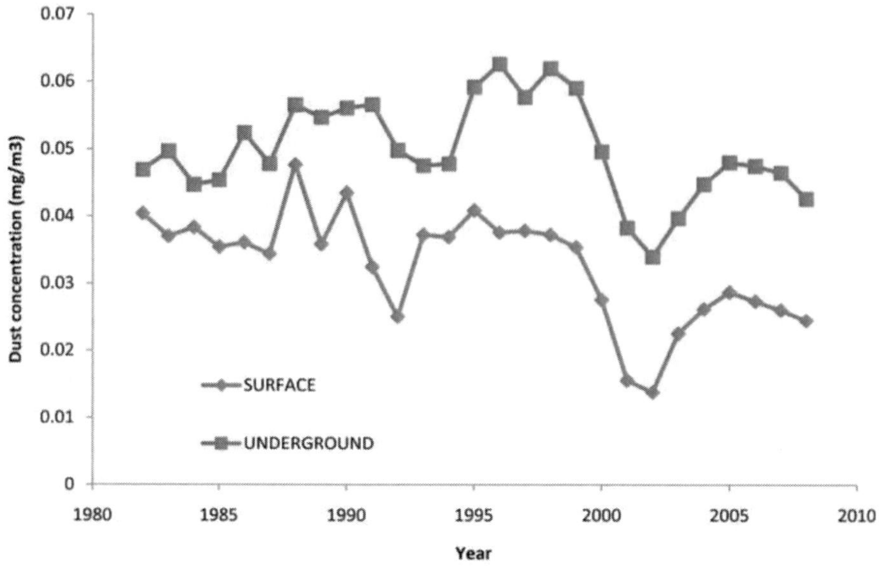

Figure 18. Respirable quartz dust: Geometric mean exposures by type of coal mine (operator and inspector data combined). [MSHA coal mine inspector and mine operator dust data].

4. Cancer Outcomes

Two cancer outcomes—lung cancer and stomach cancer—have been of particular interest with respect to work in coal mining. Lung cancer has been suspected to arise in coal miners because of their exposure to crystalline silica dust, which has been determined to be a Group I carcinogen by the International Agency for Research on Cancer, at least in some occupational settings [77]. However, findings in coal miners have been conflicting and have not strongly supported a relationship between coal mine dust exposure and lung cancer. The post-1995 findings continue this picture. No overall excess or relationship with increasing dust exposure was seen in lung cancer mortality in a study of U.S. underground coal miners [46]. However, this study, lacking silica dust exposure measurements, could not effectively evaluate the hypothesis of interest. In contrast, a recent British study that did include cumulative crystalline silica dust exposures found a weak relationship of silica exposure with lung cancer mortality [47]. A recent development in this regard is the finding that lung-deposited silica or coal dust inhibits the induction of cytochrome P4501A1 by polycyclic aromatic hydrocarbons (PAH) [78–80]. It is hypothesized that the resulting lower cytochrome activity might to some extent counteract the carcinogenic effects of tobacco smoke by limiting metabolism of PAH in tobacco smoke into carcinogenic metabolites. This may explain the lack of clear findings on dust exposure and lung cancer in coal mining.

There have been occasional reports of elevated stomach cancer mortality among coal miners. The post-1995 results from various reports have not confirmed these findings. In particular, no relationship was detected in the two studies having quantitative exposure measurements [46, 47].

5. Dust Exposure Levels, Control, and Compliance

5.1. Dust Exposure Levels

Overall trends in reported coal mine dust and crystalline silica exposure levels for the United States are shown in Figures 12 and 18. These follow the format of the *2007 WoRLD Surveillance Report* [23] Figures 2–6 and 3–5a, but are updated to 2008. The data in both figures imply that dust levels have declined over time, with those from recent years being about 75% of those around 1980, overall. This has occurred over a time period when underground production levels from both longwall and continuous miner operations significantly increased. However, the reductions vary depending on the type of mine, the source of the data, and the type of dust. The biggest reduction in reported levels was for coal mine dust at surface mines as sampled by inspectors (recent levels are ~40% of those around 1980). The smallest decline was for silica levels in underground mines, where there has been essentially no change over the time period (recent levels are ~98% of those in the early 1980s). Overall, levels of both coal mine dust and crystalline silica dust were reported to be higher in underground mines than in surface mines.

5.2. Dust Exposure Assessment

The primary advance since 1995 in the dust assessment arena has been the development of a continuously-measuring personal dust monitor (PDM) [81].

The PDM enables within-shift assessment of dust exposures, permitting prompt action to intervene and reduce excessive levels. Conventional practices that rely on the gravimetric assessment of dust collected on air sample filters preclude speedy remediation because the delay in obtaining results from the dust laboratory could mean that miners continue to be over-exposed before any indication of a problem is available from the laboratory results. The personal dust monitor is now a commercially available product and, as its use is adopted by mines, more timely and targeted interventions to reduce dust exposures will be possible. In 2010, the Mine Safety and Health Administration published new rules that provide for the approval and use of the PDM, in addition to the Coal Mine Dust Personal Sampler Unit, for determining the concentration of respirable dust in coal mine atmospheres [82].

5.3. Compliance Policy and Procedures

The federal policies and procedures for regulating underground coal mine dust levels have been the subject of criticism from their introduction in 1969. Since 1995 further critiques have been published [83, 84]. The first publication provided a historical review, the basic argument being that the problems were intrinsic to a process in which an industry essentially regulates itself (i.e., through performing the airborne sampling upon which citations are based). The second publication revisited an issue that was addressed in the NIOSH CCD, in which it was recommended that MSHA not apply any upward adjustment of the REL for instrument uncertainty. Information indicating that reported dust levels from mine operator sampling were systematically lower than those obtained by mine inspectors during unannounced visits to mines to measure exposures has been published by MSHA [40].

6. Surface Coal Mining

Studies published prior to the NIOSH CCD showed that U.S. surface coal miners (particularly workers on drill crews) were at risk of developing CWP (or silicosis). There was also evidence that ventilatory function was reduced in relation to the number of years worked as drill operators or helpers. Since the CCD, a British study has reported evidence of CWP among workers in the dustier jobs and an association with intensity of exposure [85]. Dust exposures were generally <1 mg/m^3. In the United States, a relationship between tenure in surface coal mining jobs and prevalence of CWP (ILO category 1/0 or greater, and PMF) was reported [16].

SUMMARY

A considerable body of literature has been produced from studies of coal miners and the coal mining environment since 1995, both in the United States and elsewhere. Many of the newer publications, particularly those from other countries, lack quantitative dust exposure measurements, prohibiting full and valid examination of exposure-response relationships. Nevertheless, their findings all support early findings on British and U.S. coal miners, reinforcing the generally-accepted understanding that exposure to coal mine dust can give rise to various respiratory diseases, and that those diseases can cause disability and premature mortality. The remainder of the newer publications that do have quantitative exposure data report findings that refine or augment the fundamental exposure-response results summarized in the CCD.

Overall, the following conclusions can be made:

1. No new findings have emerged since 1995 that contradict the basic summarization of the respiratory health effects of coal mine dust and their relationship with dust exposures described in the CCD [1].
2. No new findings have emerged that substantially modify the basic understanding of coal mine dust exposure and its impact on respiratory health described in the CCD.
3. The new findings that have emerged strengthen prior results and also refine or add further knowledge on disease patterns and etiology described in the CCD.
4. Overall, the logical basis for recommendations concerning prevention of occupational respiratory disease among coal miners remains essentially unaffected by the newer findings that have emerged since publication of the CCD.

New findings of particular note are:

1. After a long period of declining CWP prevalence, recent federal surveillance data indicate that the prevalence is rising.
2. Coal miners are developing severe CWP at relatively young ages.

3. There is some indication that the mortality of younger coal miners from CWP is increasing. These workers would have been employed all of their working lives in environmental conditions mandated by the 1969 Coal Mine Act.
4. The pattern of CWP occurrence across the nation is not uniform; hot spots of disease appear to be concentrated in the central Appalachian region of southern WV, eastern KY, and western VA.
5. The cause of this resurgence in disease is likely multifactorial. Possible explanations include excessive exposure due to increases in coal mine dust levels and duration of exposure (longer working hours), and increases in crystalline silica exposure (see below). As indicated by data on disease prevalence and severity, workers in smaller mines may be at special risk.
6. Given that the more productive seams of coal are being mined out, a transition by the industry to mining thinner coal seams and those with more rock intrusions is taking place and will likely accelerate in the future. Concomitant with this is the likelihood of increased potential for exposure to crystalline silica, and associated increased risk of silicosis, in coal mining.

In summary, every effort needs to be made to reduce exposures both to respirable coal mine dust and to respirable crystalline silica. As recommended in the CCD, the latter task requires establishing a separate compliance standard in order to effectively limit exposure to silica dust.

References

[1] National Institute for Occupational Safety and Health. (1995). *Criteria for a recommended standard: Occupational exposure to coal mine dust*. DHHS (NIOSH) Publication No. 95–106. Washington, DC, National Institute for Occupational Safety and Health.

[2] Attfield, M. D. & Wagner, G. (1992). *Respiratory disease in coal miners. In: Rom WN ed. Environmental and occupational medicine*. Boston, MA: Little, Brown and Company, 325–344.

[3] International Labour Office (1980). *International classification of radiographs of pneumoconiosis 1980 edition*. Occupational Safety and Health Series no. 22 (Rev. 80). Geneva: International Labour Office, 1–48.

[4] Jacobsen, M., Rae, S., Walton, W. H. & Rogan, J. M. (1971). *The relation between pneumoconiosis and dust exposure in British coal mines.* In: Walton WH ed. Inhaled particles III. Old Woking, England: Unwin Brothers, 903–919.

[5] Hurley, J. F. & Maclaren, W. M. (1987). *Dust-related risks of radiological changes in coalminers over a 40-year working life: Report on work commissioned by NIOSH.* TM/79/09. Edinburgh, Scotland, Institute of Occupational Medicine.

[6] Attfield, M. D. & Seixas, N. S. (1995). Prevalence of pneumoconiosis and its relationship to dust exposure in a cohort of U.S. bituminous coal miners and ex-miners. *Am J Ind Med*, *27*, 137–151.

[7] Attfield, M. D. & Morring, K. (1992). An investigation into the relationship between coal workers' pneumoconiosis and dust exposure in U.S. coal miners. *Am Ind Hyg Assoc J*, *53*, 486–492.

[8] Attfield, M. D. & Hodous, T. K. (1992). Pulmonary function of U.S. coal miners related to dust exposure estimates. *Am Rev Respir Dis*, *14*, 605–609.

[9] Marine, W. M., Gurr, D. & Jacobsen, M. (1988). Clinically important respiratory effects of dust exposure and smoking in British coal miners. *Am Rev Respir Dis*, *137*, 106–112.

[10] Soutar, C., Campbell, S., Gurr, D., Lloyd, M., Love, R., Cowie, H., Cowie, A. & Seaton, A. (1993). Important deficits of lung function in three modern colliery populations—relations with dust exposure. *Am Rev Respir Dis*, *147*, 797–803.

[11] Boehlecke, B. (1986). *Laboratory assessment of respiratory impairment for disability evaluation.* DHHS (NIOSH) Publication No. 86–102. In: Merchant JA ed. Cincinnati, OH: U.S. Department of Health and Human Services.

[12] ATS (1991). Lung function testing: selection of reference values and interpretive strategies. *Am Rev Respir Dis*, *144*, 1202–1218.

[13] Kuempel, E. D., Smith, R. J., Attfield, M. D. & Stayner, L. T. (1997). Risks of occupational respiratory diseases among U.S. coal miners. *Appl Occup Environ Hyg*, *12*, 823–831.

[14] Seixas, N. S., Robins, T. G., Attfield, M. D. & Moulton, L. H. (1993). Longitudinal and cross sectional analyses of exposure to coal mine dust and pulmonary function in new miners. *Br J Ind Med*, *50*, 929–937.

[15] U.S. Department of Labor. (1996). *Report of the Secretary of Labor's Advisory Committee on the Elimination of Pneumoconiosis among Coal Mine Workers.* Washington, DC: US Department of Labor,

[16] CDC (2003). Pneumoconiosis prevalence among working coal miners examined in federal chest radiograph surveillance programs—United States, 1996–2002. *MMWR Morb Mortal Wkly Rep*, *52*, 336–340.

[17] CDC (2006). Advanced cases of coal workers' pneumoconiosis—two counties, Virginia, 2006. *MMWR Morb Mortal Wkly Rep*, *55*, 909–913.

[18] CDC (2007). Advanced pneumoconiosis among working underground coal miners—Eastern Kentucky and Southwestern Virginia, 2006. *MMWR Morb Mortal Wkly Rep*, *56*, 652–655.

[19] Laney, A. S., Attfield, M. D. (2010). Coal workers' pneumoconiosis and progressive massive fibrosis are increasingly more prevalent among workers in small under ground coal mines in the United States. *Occup Environ Med*, *67*, 428–431.

[20] Loomis, D. (2010). Basic protections are still lacking. *Occup Environ Med*, *67*, 361.

[21] Seaton, A. (2010). Coal workers' pneumoconiosis in small mines in the United States. *Occup Environ Med*, *67*, 364.

[22] Antao, V. C., Petsonk, E. L., Sokolow, L. Z., Wolfe, A. L., Pinheiro, G. A., Hale, J. M. & Attfield, M. D. (2005). Rapidly progressive coal workers' pneumoconiosis in the United States: geographic clustering and other factors. *Occup Environ Med*, *62*, 670–674.

[23] National Institute for Occupational Safety and Health (2008). *Work-related lung disease surveillance report 2007*, Volume *1*. DHHS (NIOSH) Publication No. 2008–143a. Cincinnati, OH, National Institute for Occupational Safety and Health.

[24] National Institute for Occupational Safety and Health (2010). *Work-related lung disease (WORLD) surveillance system.* http://www2a.cdc.gov/drds/WorldReportData/

[25] Wade, W. A., Petsonk, E. L., Young, B., Mogri, I. (2011). *Severe Occupational Pneumoconiosis Among West Virginia Coal Miners: 138 Cases of Progressive Massive Fibrosis Compensated Between 2000–2009.* Chest DOI 10.1378/chest. 10-1326:1–14.

[26] Smith, D. R. & Leggat, P. A. (2006). 24 years of pneumoconiosis mortality surveillance in Australia. *J Occup Health*, *48*, 309–313.

[27] Baur, X. & Latza, U. (2005). Non-malignant occupational respiratory diseases in Germany in comparison with those of other countries. *Int Arch Occup Environ Health*, *78*, 593–602.

[28] Naidoo, R. N., Robins, T. G., Solomon, A., White, N., Franzblau, A. (2004). Radiographic outcomes among South African coal miners. *Int Arch Occup Environ Health*, *77*, 471–481.

[29] Marek, K. & Lebecki, K. (1999). Occurrence and prevention of coal miners' pneumoconiosis in Poland. *Am J Ind Med*, *36*, 610–617.

[30] Nguyen, A. L. & Matsuda, S. (1998). Pneumoconiosis problem among the Vietnamese coal mine workers. *J UOEH*, *20*, 353–360.

[31] Parihar, Y. S., Patnaik, J. P., Nema, B. K., Sahoo, G. B., Misra, I. B., Adhikary, S. (1997). Coal workers' pneumoconiosis: a study of prevalence in coal mines of eastern Madhya Pradesh and Orissa states of India. *Ind Health*, *35*, 467–473.

[32] Suarthana, E., Laney, A. S., Storey, E., Hale, J. M. & Attfield, M. D. (2011). Coal Workers' Pneumoconiosis in the United States: Regional Differences 40 Years after Implementation of the 1969 Federal Coal Mine Health and Safety Act. *Occup.Environ*. Med (In Press)

[33] Peters, R. H., Fotta, B. & Mallett, L. G. (2001). The influence of seam height on lost-time injury and fatality rates at small underground bituminous coal mines. *Appl Occup Environ Hyg*, *16*, 1028–1034.

[34] Laney, A. S., Petsonk, E. L., Attfield, M. D. (2009). Pneumoconiosis among under ground bituminous coal miners in the United States: is silicosis becoming more frequent? *Occup Environ Med*

[35] Pollock, D. E., Potts, J. O. & Joy, G. J. (2010). Investigation into dust exposures and mining practices in mines in the southern Appalachian Region. *Mining Engineering*, *62*, 44–49.

[36] Scarisbrick, D. A., Quinlan, T. R. (2002). Health surveillance for coal workers' pneumoconiosis in the United Kingdom 1998–2000. *Ann Occup Hyg*, *46* (Suppl. 1), 254–256.

[37] Brief, R. S., Scala, R. A. (1975). Occupational exposure limits for novel work schedules. *Am Ind Hyg Assoc J*, *36*, 467–469.

[38] Kenny, L. C., Hurley, F. & Warren, N. D. (2002). Estimating the risk of contracting pneumoconiosis in the UK coal mining industry. *Ann Occup Hyg*, *46*(Suppl. 1), 257–260.

[39] Boden, L. I. & Gold, M. (1984). The accuracy of self-reported regulatory data: The case of coal mine dust. *Am J Ind Med*, *6*, –440.

[40] U.S. Mine Safety and Health Administration. (1993). Report of the Statistical Task Team of the Coal Mine Respirable Dust Task Group. Washington DC: US Department of Labor,

[41] CDC (2004). Changing patterns of pneumoconiosis mortality—United States, 1968– 2000. *MMWR Morb Mortal Wkly Rep*, *53*, 627–632.

[42] CDC (2009). Coal workers' pneumoconiosis-related years of potential life lost before age 65 years—United States, 1968–2006. *MMWR Morb Mortal Wkly Rep*, *58*, 1412–1416.

[43] Starzynski, Z., Marek, K., Kujawska, A. & Szymczak, W. (1996). Mortality among coal miners with pneumoconiosis in Poland. *Int J Occup Med Environ Health*, *9*, 279–289.

[44] Yi, Q. & Zhang, Z. (1996). The survival analyses of 2738 patients with simple pneumoconiosis. *Occup Environ Med*, *53*, 129–135.

[45] Meijers, J. M., Swaen, G. M. & Slangen, J. J. (1997). Mortality of Dutch coal miners in relation to pneumoconiosis, chronic obstructive pulmonary disease, and lung function. *Occup Environ Med*, *54*, 708–713.

[46] Attfield, M. D. & Kuempel, E. D. (2008). Mortality among U.S. underground coal miners: A 23-year follow-up. *Am J Ind Med*, *51*, 231–245.

[47] Miller, B. G. & MacCalman, L. (2010). Cause-specific mortality in British coal workers and exposure to respirable dust and quartz. *Occup Environ Med*, *67*, 270–276.

[48] Dalal, N. S., Newman, J., Pack, D., Leonard, S. & Vallyathan, V. (1995). Hydroxyl radical generation by coal mine dust: Possible implication to coal workers' pneumoconiosis. *Free Radic Biol Med*, *18*, 11–20.

[49] Wallace, W. E., Keane, M. J., Harrison, J. C., Stephens, J. W., Brower, P. S., Grayson, R. L. & Attfield, M. D. (1995). *Surface properties of silica in mixed dusts*. In: Castranova V, Vallyathan V, Wallace WE eds. Silica and silica-induced lung diseases. Boca Raton: CRC Press, pp. 107–117.

[50] Huang, X., Li, W., Attfield, M. D., Nadas, A., Frenkel, K. & Finkelman, R. B. (2005). Mapping and prediction of coal workers' pneumoconiosis with bioavailable iron content in the bituminous coals. *Environ Health Perspect*, *113*, 964–968.

[51] Zhang, Q., Dai, J., Ali, A., Chen, L. & Huang, X. (2002). Roles of bioavailable iron and calcium in coal dust-induced oxidative stress: possible implications in coal workers' lung disease. *Free Radic Res*, *36*, 285–294.

[52] McCunney, R. J., Morfeld, P. & Payne, S. (2009). What component of coal causes coal workers' pneumoconiosis? *J Occup Environ Med*, *51*, 462–471.

[53] Kuempel, E. D., Attfield, M. D., Vallyathan, V., Lapp, N. L., Hale, J. M., Smith, R. J. & Castranova, V. (2003). Pulmonary inflammation and crystalline silica in respirable coal mine dust: dose-response. *J Biosci*, *28*, 61–69.

[54] Page, S. J. & Organiscak, J. A. (2000). Suggestion of a cause-and-effect relationship among coal rank, airborne dust, and incidence of workers' pneumoconiosis. *AIHAJ*, *61*, 785–787.

[55] Soutar, C. A., Hurley, J. F., Miller, B. G., Cowie, H. A. & Buchanan, D. (2004). Dust concentrations and respiratory risks in coalminers: key risk estimates from the British Pneumoconiosis Field Research. *Occup Environ Med*, *61*, 477–481.

[56] Buchanan, D, Miller, B. G. & Soutar, C. A. (2005). Quantitative relations between exposure to respirable quartz and risk of silicosis. *Occup Environ Med*, *60*, 159–164.

[57] Coggon, D. & Taylor, A. N. (1998). Coal mining and chronic obstructive pulmonary disease: a review of the evidence. *Thorax*, *53*, 398–407.

[58] Henneberger, P. K. & Attfield, M. D. (1996). Coal mine dust exposure and spirometry in experienced miners. *Am J Respir Crit Care Med*, *153*, 1560–1566.

[59] Henneberger, P. K. & Attfield, M. D. (1997). Respiratory symptoms and spirometry in experienced coal miners: Effects of both distant and recent coal mine dust exposures. *Am J Ind Med*, *32*, 268–274.

[60] Carta, P, Aru, G., Barbieri, M. T., Avata-neo, G. & Casula, D. (1996). Dust exposure, respiratory symptoms, and longitudinal decline in lung function in young coal miners. *Occup Environ Med*, *53*, 312–319.

[61] Beeckman, L. F., Wang, M. L., Petsonk, E. L., Wagner, G. R. (2001). Rapid declines in FEV_1 and subsequent respiratory symptoms, illnesses, and mortality in coal miners in the United States. *Am J Respir Crit Care Med*, *163*, 633–639.

[62] Naidoo, R. N., Robins, T. G., Becklake, M., Seixas, N. & Thompson, M. L. (2007). Cross-shift peak expiratory flow changes are unassociated with respirable coal dust exposure among South African coal miners. *Am J Ind Med*, *50*, 992–998.

[63] Naidoo, R. N., Robins, T. G., Seixas, N., Lalloo, U. G., Becklake, M. (2006). Respirable coal dust exposure and respiratory symptoms in South-African coal miners: a comparison of current and ex-miners. *J Occup Environ Med*, *48*, 581–590.

[64] Naidoo, R. N., Robins, T. G., Seixas, N., Lalloo, U. G., Becklake, M. (2005). Differential respirable dust related lung function effects between current and former South African coal miners. *Int Arch Occup Environ Health*, *78*, 293–302.

[65] Wang, M. L., Petsonk, E. L., Beeckman, L. F., Wagner, G. R. (1999). Clinically important FEV_1 declines among coal miners: an exploration of previously unrecognized determinants. *Occup Environ Med*, *56*, 837–844.

[66] Seixas, N. S., Robins, T. G., Attfield, M. D. & Moulton, L. H. (1992). Exposure-response relationships for coal mine dust and obstructive lung disease following enactment of the Federal Coal Mine Health and Safety Act of 1969. *Am J Ind Med*, *21*, 715–734.

[67] Wang, M. L., Wu, Z. E., Du, Q. G., Petsonk, E. L., Peng, K. L., Li, Y. D., Li, S. K., Han, G. H., Attfield, M. D. (2005). A prospective cohort study among new Chinese coal miners—The early pattern of lung function change. *Occup Environ Med*, *62*, 800–805.

[68] Wang, M. L., Wu, Z-E, Du, Q-G, Peng, K-L, Li, Y-D, Li, S-K, Han, G-H & Petsonk, E. L. (2007). Rapid decline in forced expiratory volume in 1 second (FEV_1) and the development of bronchitic symptoms among new Chinese coal miners. *J Occup Environ Med*, *49*, 1143–1148.

[69] Kuempel, E. D., Stayner, L. T., Attfield, M. D. & Buncher, C. R. (1995). Exposure-response analysis of mortality among coal miners in the United States. *Am J Ind Med*, *28*, 167–184.

[70] Sircar, K., Hnizdo, E., Petsonk, E. & Attfield, M. (2007). Decline in lung function and mortality: implications for medical monitoring. *Occup Environ Med*, *64*, 461–466.

[71] Naidoo, R. N., Robins, T. G. & Murray, J. (2005). Respiratory outcomes among South African coal miners at autopsy. *Am J Ind Med*, *48*, 217–224.

[72] Vallyathan, V, Green, F. H. Y., Brower, P. & Attfield, M. (1997). The role of coal mine dust exposure in the development of pulmonary emphysema. *Ann Occup Hyg*, *41*, 352–357.

[73] Kuempel, E. D., Wheeler, M. W., Smith, R. J., Vallyathan, V. & Green, F. H. (2009). Contributions of dust exposure and cigarette smoking to emphysema severity in coal miners in the United States. *Am J Respir Crit Care Med*, *180*, 257–264.

[74] Kuempel, E. D., Vallyathan, V. & Green, F. H. Y. (2009). Emphysema and pulmonary impairment in coal miners: Quantitative relationship with dust exposure and cigarette smoking. *Journal of Physics Conference Series*, *151*, 1–8.

[75] Brichet, A, Wallaert, B, Gosselin, B, Remy-Jardin, M, Voisin, C, Lafitte, J. J. & Tonnel, A. B. (1997). Primary diffuse interstitial fibrosis in coal miners: a new entity? *Rev Mal Respir*, *14*, 277–285.

[76] Brichet, A., Tonnel, A. B., Brambilla, E., Devouassoux, G., Remy-Jardin, M., Copin, M. C., Wallaert, B. (2002). Chronic interstitial pneumonia with honeycombing in coal workers. *Sarcoidosis Vasc Diffuse Lung Dis*, *19*, 211–219.

[77] International Agency for Research on Cancer. (1997). *IARC monographs on the evaluation of carcinogenic risks to humans: silica, some silicates, coal dust, and para-aramid fibrils*, pp. 337–406. Geneva, Switzerland, World Health Organization, International Agency for Research on Cancer.

[78] Battelli, L. A., Ghanem, M. M., Kashon, M. L., Barger, M., Ma, J. Y., Simokevitz, R. L., Miles, P. R., Hubbs, A. F. (2008). Crystalline silica is a negative modifier of pulmonary cytochrome P-4501A1 induction. *J Toxicol Environ Health*, *71*, 521–532.

[79] Ghanem, M. M., Batteli, L. A., Mercer, R. R., Scabilloni, J. F., Kashon, M. L., Ma, J. Y., Nath, J., Hubbs, A. F. (2006). Apoptosis and Bax expression are increased by coal dust in the polycyclic aromatic hydrocarbon-exposed lung. *Env Health Perspect*, *114*, 1367–1373.

[80] Ghanem, M. M., Porter, D., Batteli, L. A., Vallyathan, V., Kashon, M. L., Ma, J. Y., Barger, M. W., Nath, J., Castranova, V. & Hubbs, A. F. (2004). Respirable coal dust particles modify cytochrome P4501A1 (CYP1A1) expression in rat alveolar cells. *Am J Respir Cell Mol Biol*, *31*, 171–183.

[81] Page, S. J., Volkwein, J. C., Vinson, R. P., Joy, G. J., Mischler, S. E., Tuchman, D. P., McWilliams, L. J. (2008). Equivalency of a personal dust monitor to the current United States coal mine respirable dust sampler. *J Environ Monit*, *10*, 96–101.

[82] U.S. Department of Labor (2010). 30 CFR Part 74 RIN 1219-AB61 Coal Mine Dust Sampling Devices. *Federal Register*, *75*, 17512–17529.

[83] Weeks, J. L. (2003). The fox guarding the chicken coop: Monitoring exposure to respirable coal mine dust, 1969–2000. *Am J Public Health*, *93*, 1236–1244.

[84] Weeks, J. L. (2006). The Mine Safety and Health Administration's criterion threshold value policy increases miners' risk of pneumoconiosis. *Am J Ind Med*, *49*, 492–498.

[85] Love, R. G., Miller, B. G., Groat, S. K., Hagen, S., Cowie, H. A., Johnston, P. P., Hutchison, P. A., Soutar, C. A. (1997). Respiratory health effects of opencast coalmining: a cross sectional study of current workers. *Occup Environ Med*, *54*, 416–423.

INDEX

A

B

C

D

E

F

G

H

I

L

M

N

O

P

Q

R

S

T

U

V

W

X